Policies for Sustainably Managing Canada's Forests

Sustainability and the Environment

The Sustainability and the Environment series provides a comprehensive, independent, and critical evaluation of environmental and sustainability issues affecting Canada and the world today. Other volumes in the series are:

Anthony Scott, John Robinson, and David Cohen, eds., *Managing Natural Resources in British Columbia: Markets, Regulations, and Sustainable Development*

John B. Robinson, *Life in 2030: Exploring a Sustainable Future for Canada*

Ann Dale and John B. Robinson, eds., *Achieving Sustainable Development*

John T. Pierce and Ann Dale, eds., *Communities, Development, and Sustainability across Canada*

Robert F. Woollard and Aleck Ostry, eds., *Fatal Consumption: Rethinking Sustainable Development*

Ann Dale, *At the Edge: Sustainable Development in the 21st Century*

Mark Jaccard, John Nyboer, and Bryn Sadownik, *The Cost of Climate Policy*

Glen Filson, ed., *Intensive Agriculture and Sustainability: A Farming Systems Analysis*

Mike Carr, *Bioregionalism and Civil Society: Democratic Challenges to Corporate Globalism*

Ann Dale and Jenny Onyx, eds., *A Dynamic Balance: Social Capital and Sustainable Community Development*

Ray Côté, James Tansey, and Ann Dale, eds., *Linking Industry and Ecology: A Question of Design*

Glen Toner, ed., *Sustainable Production: Building Canadian Capacity*

Ellen Wall, Barry Smit, and Johanna Wandel, eds., *Farming in a Changing Climate: Agricultural Adaptation in Canada*

Derek Armitage, Fikret Berkes, Nancy Doubleday, eds., *Adaptive Co-Management: Collaboration, Learning, and Multi-Level Governance*

SUSTAINABILITY
AND THE
ENVIRONMENT

Policies for Sustainably Managing Canada's Forests

Tenure, Stumpage Fees, and
Forest Practices

MARTIN K. LUCKERT, DAVID HALEY,
and GEORGE HOBERG

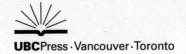

UBCPress · Vancouver · Toronto

21 20 19 18 17 16 15 14 13 12 11   5 4 3 2 1

Printed in Canada on FSC-certified ancient-forest-free paper
(100% post-consumer recycled) that is processed chlorine- and acid-free.

Library and Archives Canada Cataloguing in Publication

Luckert, Martin Karl, 1961-
    Policies for sustainably managing Canada's forests : tenure, stumpage fees, and forest practices / Martin K. Lucker, David Haley, and George Hoberg.

(Sustainability and the environment, ISSN 1196-8575)
Includes bibliographical references and index.
Also issued in electronic format.
ISBN 978-0-7748-2066-0

    1. Sustainable forestry – Canada. 2. Forest policy – Canada. 3. Forest management – Canada. 4. Land tenure – Canada. I. Haley, David II. Hoberg, George, 1958- III. Title. IV. Series: Sustainability and the environment

SD145.L83 2011   333.750971   C2011-902362-8

---

e-book ISBNs: 978-0-7748-2068-4 (pdf); 978-0-7748-2069-1 (epub)

Canadä

UBC Press gratefully acknowledges the financial support for our publishing program of the Government of Canada (through the Canada Book Fund), the Canada Council for the Arts, and the British Columbia Arts Council.

This book has been published with the help of a grant from the Canadian Federation for the Humanities and Social Sciences, through the Aid to Scholarly Publications Program, using funds provided by the Social Sciences and Humanities Research Council of Canada, and with the help of the K.D. Srivastava Fund.

UBC Press
The University of British Columbia
2029 West Mall
Vancouver, BC V6T 1Z2
www.ubcpress.ca

Contents

Illustrations

Acknowledgments

Funding for this work was provided by the Sustainable Forest Management Network. Thanks to the many provincial representatives who provided information about their tenure systems. Thanks also to the reviewers for their comments.

Policies for
Sustainably Managing
Canada's Forests

Introduction

Public forest policies in Canada matter. Canada's forests cover over 41 percent of the country's land area (Canadian Forest Service 2007), account for about 8 percent of the world's forest area (FAO 2005), and are integral to the country's history, culture, economy, and environment. Forests are the traditional home to a majority of Canada's Aboriginal people, who, for thousands of years before the arrival of European settlers, relied on them for food, clothing, shelter, medicines, material for tools and crafts, and as a source of spiritual inspiration.

Since the earliest days of colonization, forest products have been the economic mainstay of local, regional, provincial, and national economies. Today, the forest sector remains an essential component of the country's economy. Forest products, although challenged by the increasing importance of energy in recent years, remain a critical component of Canada's international trade and a major source of foreign exchange. In 2006, forest products exports were valued at Cdn$38.2 billion and contributed Cdn$28.1 billion (i.e., 55 percent) to the nation's total trade balance (Canadian Forest Service 2007). Canada is the world's largest producer of newsprint and second largest producer of wood pulp and softwood lumber. It contributes more to international trade in these three commodities than any other country (Canadian Forest Service 2006).

Although, in 2006, the forest sector accounted for only 2 percent of total employment in Canada and contributed a modest 2.9 percent of the nation's gross domestic product (Canadian Forest Service 2007), every province in Canada, with the exception of Prince Edward Island, has a sizable forest products industry that makes a significant contribution to the provincial economy, and in several provinces, most notably British Columbia, Quebec, and New Brunswick, forest products dominate the manufacturing sector. The forest sector is particularly important to the economic and social well-being of rural Canada. A study based on 1996 data of 3,853 rural census sub-districts across the country showed that forest sector employment accounted for 10 percent or more of the total labour force in 24 percent of them and 20 percent or more in 10 percent (Stedman, Parkinson, and Beckley 2005). In 2006, the Canadian Forest Service reported that over three hundred rural communities across Canada depend on the forest industry for at least 50 percent of their income (Canadian Forest Service 2006).

In addition to their importance as a source of wood products, forests produce many other products and services that have a positive impact on the lives of millions of Canadians. But in most cases, these resources do not have a value assigned through conventional market mechanisms and are invisible in the national accounts. These include supplies of high-quality water, flood control, protection from landslides and avalanches, soil conservation, micro-climatic modification, biodiversity, outdoor recreation in its myriad forms, food from forest dwelling animals and plants, and an agreeable environment in which to live and work. Furthermore, Canada's forests play an important role in the global carbon cycle and, under the Kyoto Protocol, accounting for carbon sequestration may play an important role in Canadian forest management. It would not be an exaggeration to say that forests touch the lives of almost all Canadians in a significant and positive way.

Since 93 percent of Canada's 402 million hectares of forestland is publicly owned – 77 percent provincially and 16 percent federally (see Figure I.1) – public policies have a significant impact on our forests' health, conservation, management, and the mix of goods and services they provide. Public forest policies also shape the structure and performance of Canada's wood products manufacturing sector. Since

Figure I.1   Forest land ownership in Canada

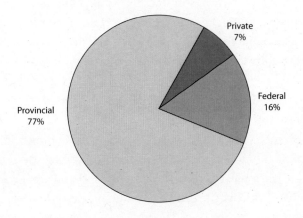

Private
7%

Federal
16%

Provincial
77%

Source: Canadian Forest Service 2007

Confederation, Canadian governments have used forest policy as a means of pursuing social, economic, and environmental goals, including regional development, job creation, community stability, the conservation of forest resources, and ecosystem protection. Consequently, forest policies provide a vital direct link between federal and provincial governments and the welfare of all Canadians, particularly those living in rural areas.

Although forest policies differ substantially among the provinces, they have much in common. Forest policy makers across the country face similar challenges when formulating effective strategies to deal with evolving domestic and global economic environments, changing societal attitudes and expectations concerning forests, and the rapidly expanding knowledge base relating to the management of forest and associated ecosystems. In spite of these similarities, no contemporary nationwide descriptions and/or comparative analyses of Canadian forest policies are available, and within individual provinces, even among senior forest professionals and administrators, there is very little understanding of forest policies in other Canadian jurisdictions.[1] Consequently, the primary justification for this book is to fill an important void in the Canadian forestry and public policy literature by making available a work that describes, in some detail, provincial forest policies, with particular emphasis on Crown forest tenure arrangements; compares them in a systematic manner; and analyzes their effectiveness in achieving the goals of sustainable forest management – the overarching objective of forest policy throughout the country.

Public Forest Policy in Canada and Abroad

Public forestland ownership has been firmly entrenched as a Canadian institution since the late nineteenth century. In keeping with contemporary colonial practice, when Great Britain assumed sovereignty over territories within what is now Canada, all land and resources became the property of the British Crown.[2] Private property was established by transferring rights from the Crown to groups and individuals. These lands became known as "Crown grants," a term that is still used today, although these holdings are now more frequently referred to as "private lands." Although considerable areas of forestland were Crown-granted in the Maritime colonies and in the settled, southerly regions of Upper and Lower Canada, vast areas of Crown forests remained.

Pre-Confederation governments, faced with small populations, meagre budgets, and valuable public timber resources begging development, devised forest licensing agreements that, by attracting the private sector capital necessary for the exploitation of timber resources, would increase public revenues and stimulate economic development. Under these arrangements, harvesting rights to public timber were transferred to the private sector in return for payments to the Crown, the land itself remaining in the public domain. The first arrangements of this type were initiated in New Brunswick in the 1830s (Wynn 1980) and were adopted by the United Provinces in 1846 (Gillis and Roach 1986). In 1865, a land ordinance in the Colony of Vancouver Island authorized the granting of timber leases, initially of unlimited size and duration – a model that was adopted by British Columbia following unification of the colonies in 1866 (Pearse 1976).

When the Canadian Confederation was founded, the Constitution Act (1867) granted jurisdiction over most lands and resources, including forests, to provincial governments.[3] These arrangements were confirmed in the Constitution Act (1982).[4] Today, federal jurisdiction over forest-lands and forest sector activities in the provinces is limited to reserved federal lands, including national parks, Indian reserves, and national defence establishments; external and interprovincial trade and commerce; matters involving migratory aquatic and terrestrial species that cross provincial boundaries; the authority to make and enforce international treaties; and general powers to make laws for peace, order, and good government of Canada (Howlett 2001b).

Following Confederation, legislative initiatives toward forests were taken by the Canadian provinces, and by the federal government in the territories still under its jurisdiction. Through the creation of forest reserves, the principle of public forestland ownership was affirmed and Crown forestland was protected from non-forest uses. Subsequently, all provinces followed the pattern established prior to Confederation and put in place licensing arrangements that delegated responsibility for managing public forestland to the private sector. In return for exclusive timber harvesting rights, licence holders contributed to Crown revenues through the payment of royalties, stumpages, land rents, and other levies, and assumed varying degrees of responsibility for forest management. These arrangements became known as "Crown forest tenures."

Today, with the exception of New Brunswick, Nova Scotia, and Prince Edward Island, the majority of Canada's forestland is in public ownership (Table I.1), and most of the timber cut on these lands is harvested by private sector companies holding Crown forest tenures.

The proportion of publicly owned forestland in Canada far exceeds that of any other developed nation with a substantial forest estate (Table I.2). Furthermore, unlike Canada, where under most tenure arrangements forest management is delegated to the private sector, the majority of countries have a public agency or public corporation whose responsibility is to manage publicly owned forests. This may take the form of a government department – for example, the United States Forest Service (USFS) or the French National Forest Office (ONF), a quasi-autonomous commission such as the British Forestry Commission, or a public corporation such as the Government Trading Enterprises (GTEs) in Australia or the Landesbetriebe in certain German States (Haley and Nelson 2006). Such

Table I.1 Ownership of forestland in Canada by province and territory, 2004

Province	Provincial %	Federal %	Private %
British Columbia	95	1	4
Alberta	89	8	3
Saskatchewan	90	4	6
Manitoba	95	2	3
Ontario	91	1	8
Quebec	89	0	11
New Brunswick	48	2	50
Nova Scotia	29	3	68
Prince Edward Island	8	1	91
Newfoundland and Labrador	99	0	1
Yukon	0	100	0
Northwest Territories	0	100	0
Canada	77	16	7

Source: Canadian Forest Service 2005

organizations are responsible for most aspects of public forestland management, including the production of timber. This timber is normally sold in the form of logs, either at the mill or at the roadside – a common practice in European countries – or as standing trees. However, in some countries, for example Chile and New Zealand, commercial timber production is largely confined to privately owned lands, the public lands for the most part being managed for environmental protection and the provision of recreational services and other non-timber forest products.

The Evolution of Canadian Forest Tenure Systems as Instruments of Forest Policy

Although a majority of forestland in Canada is publicly owned, the capital necessary to harvest and process timber resources is almost

Table I.2 Forestland ownership in selected developed countries

Country	Forestland as proportion of total land area (%)	Publicly owned forestland %
Australia	24.2	72.0
Brazil	57.2	77.0
Canada	43.6	93.4
Chile	20.6	24.2
Finland	73.9	32.1
France	28.3	26.0
Germany	31.7	52.8
Japan	68.2	41.8
New Zealand	31.0	63.4
Portugal	41.3	7.3
Sweden	66.9	19.7
United Kingdom	11.8	34.2
United States	33.1	37.8

Source: Haley and Nelson 2006

entirely in private hands. Given this dichotomy, some of the most important questions facing public forest policy makers are how to transfer rights to utilize forest resources from the public to the private sector; how to capture for the public owners of the forests an equitable financial return for the use of public resources; and how to ensure that when private firms use public forest resources, broad public interests are protected and public goals are achieved.

Canadian governments, both federal and provincial, have addressed these questions through the Crown forest tenure system, which, since the earliest days of forest management, has been the principal instrument of forest policy in all jurisdictions and, consequently, is the major focus of this investigation. As described above, the tenure system is used to transfer rights to use public forests, principally the right to harvest timber, to the private sector. Financial returns to governments for the use of public resources are collected in various ways and are a component of the contractual agreements under which timber harvesting rights are transferred. The principal payments are "stumpage fees," the term used to describe the direct cost to licensees for the Crown timber they harvest (see Chapter 5). In addition to fiscal measures, governments protect the public interest in Crown forests by means of regulations that set conditions – for example, reforestation or harvesting practices – that tenure holders must fulfill in order to exercise their rights.

The history of public forest policy in the Canadian provinces has largely been a story of Crown forest tenure arrangements evolving to accommodate changing public attitudes toward forests and meet new demands on forest resources. Thus, with the passage of time, Crown forest tenures have become increasingly complex.

The early tenure arrangements simply required licensees to pay levies such as timber royalties, land rents, and renewal fees. Although some conservationists in Canada raised alarms about the future of commercial timber supplies in the late nineteenth century, there was little concern for the rate of timber harvesting, the restocking of harvested stands, or the environmental consequences of harvesting activities. Populations were small, forests seemed to be endless, and the goal of forest policy was to convert standing timber into other forms of capital, such as infrastructure for transportation and communication, as efficiently as possible to fuel developing economies that were almost entirely dependent on natural resources. There was one environmental "problem," however, that demanded the attention of forest administrators: fire. Although wildfire is a natural component of many of Canada's forest ecosystems, it was regarded by early public policy makers, as it is to a large extent today, as a destructive agent that destroys valuable timber and threatens forest-based communities. Fire prevention and suppression, therefore, became the key to early expressions of forest management in Canada and the first management responsibilities imposed on tenure holders. This was followed by a more general concern for forest protection that included insect and disease control.

By the mid-twentieth century, provincial governments faced three principal concerns. A large and increasingly capital-intensive forest industry sought more secure supplies of timber for further expansion and consolidation, foresters feared the consequences for future timber supplies of an unbalanced pattern of harvesting and inadequate reforestation, and rural populations sought greater security than was provided by an industry that advanced across the landscape leaving moribund communities in its wake.[5] To address these problems, new tenure arrangements were developed in many Canadian provinces that provided their holders with relatively long-term secure timber supplies at administratively determined stumpage prices. In return, tenure holders undertook to manage their licences for a sustained yield of timber. To a large extent, contemporary Canadian forest tenures have their roots in these mid-twentieth-century Crown forest policy reforms.

During the post-Second World War years through to the early 1970s, the major priorities of forest policy in most Canadian provinces was sustained economic growth and development, the creation of regional employment, and the mitigation of instability in forest-dependent rural communities. Tenure arrangements were designed to attract capital investment to efficiently harvest and process public timber resources in an orderly and sustainable manner. These policies, by and large, were successful in meeting their objectives. Vast areas of primary forest were opened up, new and established communities flourished, real wages in the forest industry rose as labour productivity increased, and government revenues burgeoned. By the late 1970s, however, the global and domestic environments in which the Canadian forest sector functions began to enter a period of profound change. The pace of change accelerated throughout the 1980s and 1990s and continued during the first decade of the new millennium.

During the late 1970s, the global environmental movement gathered momentum and in many countries began to attract more interest from mainstream citizens. In Canada, forests, as one might expect, became a major focus of environmentalists' concerns. Initially, provincial governments responded by introducing policies that placed more emphasis on the management of forests to produce a broad spectrum of non-timber products while continuing to emphasize the sustained production of commercial timber. However, by the early 1980s, there were increasing pressures from environmental NGOs for a new approach to forestry that would place more emphasis on integrated management designed to produce an optimum mix of timber and non-timber forest products while maintaining the health and integrity of forest ecosystems. A major turning point for the global environmental movement was the publication, in 1987, of the report of the World Commission on Environment and Development (Brundtland 1987). The report introduced the term "sustainable development," a concept that captured the public's imagination, stirred governments to action, and galvanized the efforts of the environmental movement worldwide.

By the mid-1990s, Canada's forests, as a result of high-profile campaigns such as the one in 1994 to protect the forests of Clayoquot Sound on Vancouver Island, had become a magnet for international environmental activism (J. Wilson 2001). Civil disobedience in the woods and international campaigns promoting consumer boycotts forced provincial governments to act and redirect forest policies toward more environmentally sensitive and socially conscious forest management.

In 1990, a House of Commons subcommittee advised Parliament that if Canada were ever to practise sustainable development successfully, it must begin in the forests (Standing Committee on Forestry and Fisheries 1990). Later that year, the Canadian Council of Forest Ministers (CCFM) sponsored the National Forestry Forum on Sustainable Development and Forest Management. The meeting was followed by a series of public forums on the management of Canada's forests, culminating in 1992 with the production of the document *Sustainable Forests: A Canadian Commitment* (CCFM 1992). Throughout the balance of the 1990s and into the new millennium, sustainable forest management became the major focus of the environmental policy agendas of both federal and provincial governments.

Although sustainable forest management has been defined in many ways (e.g., National Forest Strategy Coalition 1992; Helsinki Process 1993; International Tropical Timber Organization 1998), central to all definitions is the maintenance of the health and integrity of forest ecosystems in a condition that will allow them to flourish and produce multiple economic, social, and cultural benefits for present and future generations. Sustainable forest management, therefore, is generally regarded as having three dimensions: economic sustainability, social sustainability, and environmental sustainability. In pursuit of sustainable forest management, governments must endeavour to achieve a balance among economic goals; social goals, such as equity, job creation, and community stability; and the protection of healthy, intact forest ecosystems capable of providing consumptive and non-consumptive goods and environmental services in perpetuity.[6]

Individual provinces responded to the challenges of sustainable forest management in various ways. However, on the whole, provincial Crown tenure systems have remained essentially intact and continue to place emphasis on sustained timber harvests and the notion that regional economic prosperity and sustained regional development can be achieved by regulating the structural development, timber utilization, and marketing strategies of the forest industry. Changing social demands on forests and the ecological goals of sustainable forest management have not been accommodated by modifying forest tenures in ways that provide incentives for private sector licensees to voluntarily work toward emerging public goals. Rather, tenure holders are required by increasingly complex and pervasive regulations to provide for non-timber forest products, environmental services, and ecosystem protection (see Chapter 4). Such measures often result in considerable compliance and enforcement costs and may significantly attenuate licensees' timber harvesting rights (G.C. van Kooten 1994; Haley 1996).

Since the mid-1990s, the ability of Canada's pulp and paper and lumber manufacturing sectors to compete in global markets has been severely eroded, resulting in declining market shares; falling rates of return to capital; declining capital investment that, in some parts of the country, is insufficient to even replace depreciated capital stocks; and an increasing number of mill closures, accompanied by falling employment and declining local economies.[7] In conjunction with increasing costs from environmental requirements of tenure policies, a number of other factors, both domestic and global, are responsible for this situation that threatens to undermine the important historic role the forest sector has played in the Canadian economy. Among the domestic factors are declining supplies of economically accessible timber, deteriorating timber quality, an aging and difficult-to-replace labour force, and rising energy costs. Global factors include a realignment of global currency exchange rates, resulting in a rising Canadian dollar relative to most other currencies; increased capital mobility; the ongoing and protracted softwood lumber trade dispute with Canada's largest customer, the United States; a realignment of global supply/consumption relationships as new, low-cost suppliers emerge and consumers switch to more readily available and less costly lower-grade wood products and non-wood substitutes; the emergence of China as both a major market for, and competitive global supplier of, wood products; and a continuing secular downward trend in real prices for both pulp and lumber for the foreseeable future (Forest Economics and Policy Forum 2005). These trends were exacerbated in the late summer of 2008 by the onset of a major worldwide recession precipitated by the near collapse of the global financial sector.

Although all provincial governments recognize the need for forest policies that encourage global competitiveness and create a favourable environment for capital investment, in most provinces, for the past decade or more, these goals have frequently been overshadowed by policy initiatives designed to achieve environmental and other social objectives at the expense of industrial competitiveness.[8] However, recently several provinces have embarked on analyses of their forest industries and are exploring policy reforms that might be introduced in order to reverse the sector's downward economic spiral. In November 2004, the Ontario minister of natural resources established the Minister's Council on Forest Sector Competitiveness, which brought down its final report in June 2005 (Minister's Council on Forest Sector Competitiveness 2005). On 30 March 2005, British Columbia's premier announced the formation of the BC Competition Council to review the province's competitive position and to recommend workable private sector and public sector actions to improve British Columbia's competitiveness (BC Office of the Premier 2005). Important components of the council were the Pulp and Paper Industry Advisory Committee and Wood Products Industry Advisory Committee. The former brought down its Final Report on 25 January 2006 (BC Competition Council 2006a), and the latter did so on 31 March 2006 (BC Competition Council 2006b). In 2008, the BC minister of forests and range appointed a multistakeholder working round table on forestry (BC Ministry of Forests

and Range 2008) "to develop recommendations and ideas that government, the forestry industry and others can act on to strengthen British
Columbia's forest sector over the next five to 10 years." In Alberta, the
provincial government in cooperation with the Alberta Forest Products
Association prepared a report on the economic health of the Alberta
forest industry and its ability to compete in the global marketplace. This
report was followed by the establishment of the ongoing Forest Industry
Sustainability Committee, with representation from industry and government ministers.[9] It is likely that these and similar initiatives that are
being contemplated in other provinces will result in reforms to existing
forest tenure arrangements that place more emphasis than in recent
years on economic sustainability.

Objectives and Organization of This Book

As is evident from the preceding discussion, Crown forest tenures are,
and have been historically, a major instrument used for the implementation of public forest policies in Canada. In fact, if a comprehensive approach is taken to the investigation of tenure arrangements that includes
the rights they grant and the broad spectrum of constraints they place
on their holders' behaviour, then many significant aspects of forest policy
are embraced. Thus, the principle objective of this book is to describe
Canada's provincial Crown forest tenure systems within analytical
frameworks that allow us to compare alternative policy approaches and
analyze their suitability for pursuing sustainable forest management.
Each of the chapters in this book is targeted toward making a contribution to this overarching objective.

   In Chapter 1, we describe how the emergence of sustainable forest
management is occurring in concert with a number of key changes in
the role of the state in governance. Forest tenure arrangements determine
how the responsibility for delivering the many products and environmental services provided by public forests is allocated between provincial
governments and other agents – private firms and individuals, First
Nations, local authorities, communities, NGOs, and other stakeholders.
Forest tenures, therefore, play a central role in forest sector governance.
But forest governance and sustainable forest management transcend
considerations of forest tenures and include such approaches as forest

certification and criteria and indicators. Chapter 1 provides this broader context for the discussions that follow by examining the changing roles of the state in relation to private markets and civil society, both generally and in relation to the rise of sustainable forest management within the Canadian forest sector. The emergence of new governance models in the forest sector are examined, including the evolving roles of First Nations, environmental NGOs, communities, stakeholder organizations, and private certification bodies.

To describe and analyze Canada's complex Crown forest tenure systems in a way that allows interprovincial comparisons to be made and their outcomes to be critically examined requires an appropriate analytical framework. The purpose of Chapter 2 is to develop such a framework. Following a discussion of the importance of property rights as instruments of public policy, particularly their roles in market economies, Chapter 2 goes on to explain the analytical approach adopted in this study. This methodology, which is sometimes referred to as the "property rights approach" to public policy analysis (Alchian and Demetz 1973; Bromley 1991), is predicated on the notion that property rights can be disaggregated into a number of attributes or characteristics. Any property right can be perceived as a combination of "rights" that authorize its holder to extract a stream of benefits, or utility, from the property concerned and concomitant obligations that "attenuate" the value of this benefit stream. Property rights characteristics describe both rights and obligations. Thus, each forest tenure can be described in terms of its characteristics that identify its unique features, facilitate comparisons with other tenure types, and, taken together, provide a powerful tool that can assist policy analysts in understanding the relationships between different tenure arrangements and their possible outcomes as reflected in the behaviour of their holders. The system of property rights characteristics adopted for this study is described, and the possible impact of individual characteristics on the behaviour of forest tenure holders is discussed.

The property rights embodied in forest tenures in Canada are sufficiently complex to justify the use of multiple analytical frameworks. Two characteristics of Crown forest tenures – regulations designed to control forest practices and Crown stumpage fees – are so complex and such important components of contemporary forest policies throughout Canada that they receive special attention in this study. Each is the subject of a separate chapter (Chapters 4 and 5 respectively), where special analytical frameworks are developed and applied to describe and classify these important sources of attenuations in ways that allow useful interprovincial comparisons to be made.

In Chapters 3 through 5, we populate our analytical frameworks with information about, respectively, forest tenures, regulations, and stumpage fee systems across Canada. Chapter 3 uses the methodology developed in Chapter 2 to describe and compare forest tenures in nine of Canada's ten provinces. Prince Edward Island is excluded, since more than 90 percent of that province's forestland is privately owned and the forest sector is not a very important component of the provincial economy. In Chapter 4, an analytical framework is developed that recognizes four approaches to regulating forest practices within four general areas: strategic planning, operational planning, requirements for specific practices, and compliance and enforcement regimes. Unlike the more general analysis of Crown forest tenures presented in Chapter 3, comparisons of forest practices regulations are confined to five provinces where the forest sector plays a fairly prominent role in the provincial economy: British Columbia, Alberta, Ontario, Quebec, and New Brunswick. In Chapter 5, following a theoretical discussion of economic rent, stumpage, and the desirable characteristics of a public stumpage system, a framework for comparing stumpage fee systems is developed based on seven distinct characteristics. This scheme is then used to describe methods of determining, collecting, and dispersing stumpage fees in the Canadian provinces, again excluding Prince Edward Island.

One of the key purposes of Chapters 3 through 5 is to provide the empirical basis for making comparisons between forest tenures across Canada. Moreover, this overview of existing tenure policies provides insights regarding how current policies may facilitate, or stand in the way of, sustainable forest management objectives. Along these lines, at the end of each of these three chapters, we identify features of current tenures, forest practices, and stumpage fee systems that may impede or facilitate the pursuit of sustainable forest management objectives. These sections provide information that foreshadows some of the conclusions reached in Chapter 6.

Chapter 6 synthesizes the information and concepts presented in previous chapters. After reviewing some of the challenges facing the forest products industry in Canada, the chapter links a number of these problems to current tenure configurations. Most of these observations are drawn from Chapters 3 through 5, where potential problems with current policies are identified. However, taken together, these problems suggest broader directions for potential change. From these potential future directions, several essential attributes are identified that may be used to guide tenure reform. The chapter then considers numerous barriers that could prevent change in Canadian forest tenure systems. We conclude by discussing potential means of overcoming these barriers.

The empirical work in this volume is based on a province-by-province survey of forest legislation, regulations, licensing contracts, and other policy documents. In the Appendix we provide province-by-province sources that were used to collect this information. In many cases, changes in forest policies were underway as the data were being collected. Thus, the empirical information presented here represents policies as they existed at a snapshot in time. Although forest tenures are in constant flux, the general structures of forest tenures provide a basis to assess general policy approaches that have been adopted, and to assess potential future directions for forest tenures.

The Rise of Sustainable Forest Management and Trends in Forest Sector Governance

1

This book is concerned with institutions and institutional change in the Canadian forest sector, specifically provincial Crown forest tenure systems. As we suggest in the Introduction, Crown forest tenures are the principal instruments used by governments in Canada to implement public forest policy. A salient function of forest tenure systems is that they serve to allocate responsibilities for the management of public forestland between governments and other agents such as private firms, First Nations, and communities. Crown forest tenures, therefore, play a central role in forest sector governance – who owns forests, who has the authority to determine how they are managed, and who is actually responsible for their management.

The forest sector has experienced dramatic changes in governance structures over the past decade. Some of these changes are sector specific, but some of them reflect broader societal trends. Broadly speaking, as the sector has grappled with implementing the concept of sustainable forest management, the magnitude and complexity of what society is demanding of its forests have increased exponentially and the traditional hierarchical systems of governance are being challenged from a variety of sources. In this chapter, trends in the role of the state in relation to markets and civil society are examined, both in general and with respect to the rise of sustainable forest management within the forest sector, including expanding participation in decisions concerning the stew-ardship of forest resources for a broader range of interest groups, the challenges of developing new relationships with First Nations, the consideration of innovative policy instruments, the emergence of corporate social responsibility, the rise of forest certification, and increasing pressures to offload management responsibilities from the government to Crown forest tenure holders.

General Trends in the Roles of Government: From Government to Governance

The relationship between the state and the market in developed economies has changed significantly over time, in measurable and ideological terms. After the steady trend of growth in government as a component of the economy in the period after the Second World War, pressures for the downsizing of government emerged in the 1980s, as witnessed by the governments of Thatcher in Britain, Reagan in the United States, and Mulroney in Canada. Even Bill Clinton, in 1992, the only Democratic president to be elected in the United States between 1980 and 2008, famously acknowledged in his 1996 State of the Union address that "the era of big government is over." Throughout the Western industrialized world, government programs have experienced budget cutbacks, and attention has shifted from relying on governments to fix "market failures" to diagnosing and addressing "government failures" (Wolf 1979; Weimer and Vining 2005). Governments and policy analysts have begun to consider a broader, more market-oriented range of policy instruments for the delivery of services traditionally performed by government.

When the "great recession" struck in the fall of 2008, governments around the world responded with a massive increase in the role of the government, including the outright purchase of major financial and industrial corporations. Although it is premature to speculate on the practical and ideological consequences of this extraordinary turn of events, it is striking that, thus far, these massive changes have generally been rationalized as a response to an emergency rather than as the basis for a new public philosophy on the role of the state. Perhaps the most profound indication of how our conceptualization of the appropriate role of government has changed is the shift in discourse from a focus on *government* to one on *governance*. In what Salamon (2002) referred to as "the revolution that no one noticed," two fundamental changes have occurred. First, there has been a marked shift toward less hierarchical, more collaborative relationships between government and society as governments have increasingly recognized how much they rely on non-governmental actors for the effectiveness and legitimacy of policy development and implementation. Salamon calls this a shift from hierarchy to network. Second, there is an increasing reliance on third parties and public-private partnerships in service delivery. As a result, much of modern governance involves the management of effective relationships among governments, non-governmental organizations, and markets (Salamon 2002; Pierre and Peters 2000).

Components of this new governance have been evident in the Canadian forest sector for some time now. Much of the current emphasis is on contracting out the delivery of services, but Canadian forest tenures have been a form of alternative service delivery since governments first required licensees to take on management responsibilities in return for exclusive rights to harvest public timber. In fact, close "bipartite" relationships have existed between governments and the forest industry for decades (Howlett and Rayner 2001).

However, much of what is involved in the "new governance" is also new to the forest sector. Over the past two decades, a wide range of new or newly strengthened interests have come to challenge the legitimacy of government-forest industry relationships (Pearse 1998; Stanbury and Vertinsky 1998).[1] Environmental groups have used both grassroots activism and international market campaigns to force private firms and governments to pay greater attention to environmental concerns. First Nations have challenged the very basis of government ownership of "Crown" land. Community groups have demanded decentralization of control, and a broad range of individuals and groups, both commercial and non-commercial, affected by forest management decisions have insisted on a larger role in the formulation and implementation of managerial strategies. The Canadian forest sector has witnessed a shift from a closed, government-industry partnership to a more open network best characterized as "multipartite bargaining." From national round tables to regional land use planning in the provinces, multi-stakeholder negotiations have become a central component of policy development (Lindquist and Wellstead 2001). The forest sector has also witnessed the emergence of non-governmental, private certification organizations that have come to share in the governance of forest management. Many of these changes are associated with the rise of sustainable forest management as a policy paradigm.

The Emergence of Sustainable Forest Management as a Policy Paradigm

Sustainable forest management emerged as the dominant paradigm for the management of forest resources in Canada in the 1990s. The concept has its roots in earlier efforts to expand the objectives of forest management beyond the focus on sustained yield that dominated the development of industrial forestry in Canada (Luckert and Williamson 2005). The concept emerged in the early 1990s as Canada prepared for the 1992 UN Conference on Environment and Development and sought to articulate a model that could provide a foundation for a legally binding international forest convention but also provide some defence against growing international environmental criticisms of Canada's forest practices (Bernstein and Cashore 2000). When the Canadian Council of Forest Ministers (CCFM) issued the report *Sustainable Forests: A Canadian Commitment* in 1992, the concept of sustainable forest management was granted official status in the discourse of Canadian forest management and policy (Rayner and Howlett 2007).

How sustainable forest management came to be defined in operational terms, and whether it has produced significant changes in forest policy and management, are challenging questions. Although definitions have varied over time and source, one broad definition that reflects the core spirit of sustainable forest management is the vision statement of the 1992 Canada Forest Accord: "Our goal is to maintain the long-term health of Canada's forest ecosystems, for the benefit of all living things, and for the social, cultural, environmental and economic well-being of all Canadians now and in the future" (as cited in Burton et al. 2003, 49). Arguably, the most fundamental departure from the prior sustained yield paradigm is the elevation of the importance of environmental and social values.

Operationally, sustainable forest management was given meaning through the articulation of *criteria* that reflect core forest values and through *indictors* that measure progress toward desired objectives (Luckert and Boxall 2009). Although the concept has been enthusiastically developed and promoted through national, interjurisdictional process through the CCFM, provincial governments still dominate jurisdiction over forest management across the country. The remainder of this chapter draws those links between new governance trends and sustainable forest management.

The Rise of the Organized Environmental Movement

The environmental component of the sustainable forest management paradigm was initially forced on the forest sector by external mobilization. One of the most important trends over the past several decades is the development and institutionalization of environmental groups active on forestry issues throughout Canada. These groups have provided a persistent challenge to the sustained yield, industrial model of forest management. Much of the initial focus was on setting aside more wilderness as protected areas, but they have become increasingly active in pressuring governments on a broader range of policies, especially forest practices. Unfortunately, there is not much literature on the Canadian environmental movement relevant to the forest sector.[2]

J. Wilson (2002) stressed three important facts about the Canadian environmental movement. First, it is extremely diverse in its organization and strategies, which we will discuss in more detail below. Second, it draws great strength from the personal commitment of the activists who staff the organizations. Third, the movement is still relatively resource-poor in terms of finances and professional staff, limiting its ability to compete successfully with government and industry organizations.

A wide variety of environmental groups address forestry issues across the country. Table 1.1 provides a sample of some of the most prominent groups nationally and in selected jurisdictions. Groups operate at various scales. Some are very large international organizations, such as Greenpeace, that have Canadian chapters. Others, such as the Sierra Club of Canada and the Canadian Parks and Wilderness Society (CPAWS), are national groups that work at the federal and national levels

Table 1.1   Prominent environmental groups focusing on forest-related issues in selected Canadian jurisdictions

Jurisdiction	Environmental group	Date founded
International (strong Canadian presence)	• ForestEthics	1994
	• Greenpeace	1971
	• Rainforest Action Network	1985
National/ federal	• Sierra Club Canada*	1963 – started activities in Canada 1989 – established office in Ottawa
	• Canadian Parks and Wilderness Society (CPAWS)*	1963
	• Canadian Boreal Initiative	2003
	• Sierra Legal Defence Fund	1990
	• World Wildlife Fund Canada	1967
	• Nature Canada (formally Canadian Nature Federation, which was formally the Audubon Society of Canada)	1939 – magazine begins 1948 – Audubon Society established
	• Canadian Wildlife Federation	1962
British Columbia	• Sierra Club of BC Foundation	1969 – started activities
	• Western Canada Wilderness Committee	1980
	• West Coast Environmental Law	1974
	• David Suzuki Foundation	1990
Alberta	• Alberta Centre for Boreal Studies (started by CPAWS Edmonton chapter)	2000
	• Alberta Wilderness Association	1965
	• Forest Watch Alberta (part of Global Forest Watch)	1999
Ontario	• Wildlands League (Ontario CPAWS chapter)	1968

▶

◀   Table 1.1

Jurisdiction	Environmental group	Date founded
	• Ontario Nature (Federation of Ontario Naturalists)	1931
	• Earthroots	1986
	• Canadian Environmental Law Association (land use issues, among others)	1970
Quebec	• Union québécoise pour la conservation de la nature (UQCN)	1981
	• Société pour la nature et les parcs du Canada – Montreal section, Quebec chapter (SNAP = CPAWS)	2001
	• Réseau québécois des groupes écologistes (RQGE)	1982
	• Aux arbres citoyens! (Combination of four groups: WWF, SNAP, UQCN, RQGE. Deals specifically with boreal forest and protected areas issues.)	2001
	• L'action boréale de l'Abitibi-Témiscamingue (vice-president is singer/songwriter Richard Desjardins, who made the documentary *l'Erreur Boréale*, which shocked Quebec and started a big forest movement)	2000
New Brunswick	• Conservation Council of New Brunswick	1969
	• Crown Lands Network (A loosely knit assemblage of New Brunswick environmental groups, such as the Conservation Council of New Brunswick, the Sierra Club Canada Atlantic chapter, and the New Brunswick Protected Natural Areas Coalition. The network is a "caucus" of the New Brunswick Environmental Network.)	2001-02

* All of the provinces listed above have a Sierra chapter (New Brunswick's is the Atlantic Sierra chapter) and a CPAWS chapter.

but also have strong provincially focused chapters. Others focus explicitly on provincial issues, and some are organized around a particular area of concern (such as Friends of Clayoquot Sound on the west coast of Vancouver Island in British Columbia). The most prominent of these groups have been active for at least ten years; some, such as Nature Canada and Greenpeace, have existed for many decades.

Environmental groups rely on a wide range of strategies to influence public policy and industrial management practices.[3] They utilize traditional group strategies, such as lobbying public officials and, more so than most groups, combine insider lobbying with efforts to mobilize public opinion. Although it has become far less common over the past five years, environmental organizations in Canadian forestry are well known for their reliance on protest and direct action to draw attention to specific causes – the protests in Clayoquot Sound in British Columbia and Temagami in Ontario being the most prominent examples (J. Wilson 1998; Cartwright 2003).

However, more frequently, media events are built around the release of reports, either focused on advocacy research reports or "report cards" evaluating government or corporate performance according to the environmental group's criteria.[4] Some examples of recent report cards are described on the following page.

Environmental groups also use litigation as a strategy. However, because of the discretionary nature of Canadian environmental law, they have not been nearly as effective as their counterparts in the United States, or First Nations in Canada, at using courts to force change. Nevertheless, environmentalists still find the occasional lawsuit effective in focusing pressure on high-priority concerns (Hoberg 2000).

The most important change in environmental group strategies has been an increasing reliance on international markets to pressure the industry and government to adopt stronger environmental policies (Bernstein and Cashore 2000; Stanbury 2000). Leading environmental groups such as Greenpeace and ForestEthics have turned their attention away from roadblocks to focus on the consumers of Canadian forest products, especially large wood products retailers like Home Depot in North America and Sainsbury's in Europe, or major publishers utilizing Canadian paper. Through persuasion backed by the threat of protests,

Recent Examples of Environmental Group Report Cards

Great Bear Rainforest Report Card
A report written by four ENGOs (Greenpeace Canada; ForestEthics; Sierra Club Canada, BC Chapter; and the Rainforest Action Network) evaluating changes in forest practices in the Great Bear Rainforest in British Columbia since land-use agreements were signed in the early 2000s. Grades range from a high of C for "use of credible science" to a low of F for "ecological management and planning."
http://www.savethegreatbear.org/resources/Reports/2005 report card

Sierra Club Canada – National Forest Strategy Report Card Database
An analysis of the provinces' fulfillments of their commitments to the National Forest Strategy.
http://www.sierraclub.ca/national/programs/biodiversity/forests/nfs/index.shtml

Sierra Club Canada – Rio Report Cards
For thirteen years, Sierra Club Canada has published report cards that grade the federal and provincial governments on their commitments made in Rio de Janeiro in 1992. The 2005 report gives Canada a C for its implementation of the forest principles.
http://www.sierraclub.ca/national/rio/

Ontario Nature (Federation of Ontario Naturalists) – Ontario Living Legacy Report Card
An assessment of the progress made in establishing parks and protected areas under Ontario's Living Legacy plan.
http://www.ontarionature.org/discover/resources/PDFs/reports/report_crd.pdf

environmental groups have persuaded a number of major companies to adopt purchasing policies that steer them away from forest products produced from companies or in areas that environmentalists have designated to be engaging in unsustainable practices. The most prominent market campaigns in Canada have been those focused on the Great Bear Rainforest, comprising the central coast region of British Columbia, and more recently the entire boreal forest region.[5]

This strategic innovation has been profound because it has created a direct corporate interest in addressing environmental issues to protect market shares. One consequence of the influence of these tactics is a dramatic increase in collaborative initiatives between leading environmental groups and the forest industry. For example, several major forest companies have signed onto the Boreal Forest Conservation Framework, which calls for protecting "at least 50% of the region in a network of large interconnected protected areas" (Canadian Boreal Initiative 2003).

Environmental interest groups have become influential players in the Canadian forest sector. They have relied on a combination of direct action, conventional interest group strategies, and innovative political strategies through the marketplace to force governments and industry to demonstrate greater protection of environmental values in the forest.

Challenges in Reconciliation with Aboriginal Peoples

Another major challenge confronting the forest sector is the increasing role of First Nations, both in the economic activities of the industry and governance of the resource. Despite the proximity of many First Nations communities to forest industry operations, First Nations have not benefited significantly from forestry in much of Canada (Wilson and Graham 2005). Over the past several decades, numerous conflicts have emerged between industrial forest development and First Nations communities seeking to protect their Aboriginal and treaty rights. Several of these cases have ended up in the courts, where the general trend has been the expansion of Aboriginal rights. A series of court decisions has forced governments and the forest industry to give greater consideration to the interests of First Nations. Although, historically, many governments and companies have been reluctant to acknowledge the need for change, there now seems to be an emerging consensus among government, industry, and First Nations on a strategy to increase First Nations' participation in the forest sector (e.g., Wilson and Graham 2005). And, although it seems somewhat slower in coming, there are increasing signs that governments have acknowledged the need to share authority with First Nations in some areas.

Several decisions by the Supreme Court of Canada have been instrumental in advancing Aboriginal rights and title. In the December 1997 *Delgamuukw* case involving the Gitksan and Wet'suwet'en people of northwestern British Columbia, the Supreme Court of Canada entrenched the principle that "there is always a duty of consultation."[6] In that case, the court elaborated on the meaning of Aboriginal title and under what circumstances it could be infringed. The court ruled that title could be infringed by the Crown for certain purposes involving the "furtherance of a legislative objective that is compelling and substantial," and explicitly included forestry operations as an example of that kind of government activity. But infringement could occur only with adequate consultation and compensation. The case had a significant impact not only on the Government of British Columbia but on the governments of other provinces as well, forcing them to adopt or revamp policies designed to improve procedures for consultations with First Nations (Ross and Smith 2002).

Despite adopting more elaborate consultation policies, the Government of British Columbia maintained in legal arguments that the duty to consult arises only once Aboriginal title has been proven, a situation that has not occurred throughout most of British Columbia. This legal position created significant tensions with First Nations groups, but the conflict was clearly resolved in the November 2004 decisions of the Supreme Court of Canada in the *Haida* and *Taku* cases. The *Haida* case involved the transfer of Tree Farm Licence 39 on the Queen Charlotte Islands/Haida Gwaii to the Weyerhaeuser Corporation when it purchased the MacMillan Bloedel Corporation in 1999; the *Taku* case involved the approval of a mining licence and associated road following a provincial environmental assessment.[7]

The *Haida* case was particularly important because lower court decisions had created a great deal of uncertainty as to how elaborate consultation needed to be, and who was responsible for consultation. In 2002, the BC Court of Appeal had extended the obligation beyond consultation to include *accommodation* of First Nations' interests. Moreover, the court expanded the legal duty of consultation and accommodation to include the licensee as well as the Crown. That decision led the

BC government to adopt new policies that included accommodation based on the development of forest and range agreements that incorporate revenue sharing with First Nations and access to volume-based timber harvesting licences in exchange for First Nations' acknowledgement that the government had satisfied the economic aspects of consultation and accommodation.

In the *Haida* decision, the Supreme Court established in no uncertain terms that the Crown has an obligation to consult *and* accommodate First Nations' interests, even if title has not been proven. The court based this ruling on the "honour of the Crown," originating from the Royal Proclamation of 1763. The court also clarified that third parties such as Weyerhaeuser do not share in that legal obligation, ruling that "the honour of the Crown cannot be delegated." At the same time, in the *Haida* and *Taku* cases, the court also clarified that the government's duty to consult and accommodate did not grant First Nations a veto over activities they oppose. In the *Taku* case, the court ruled that the elaborate process conducted by the BC government under its environmental assessment process was sufficient. These decisions do not resolve all the uncertainties about how far governments need to go to consult and accommodate First Nations, but they do clarify three important issues: existence of the duty to consult and accommodate even before title claims have been established; that this legal obligation rests with the Crown and not the licensee; and that First Nations do not hold a veto over development decisions.

Most of the focus of Aboriginal jurisprudence in Canada has been in British Columbia, where because of the absence of settled treaties there is greater uncertainty about Aboriginal rights and title. Nonetheless, First Nations issues have been increasingly prominent in the forest sector across the country. In many areas there are disputes over the meaning of existing treaties regarding the relationships between First Nations and the Crown for the control of forest resources. In New Brunswick, the Court of Appeal in 2003 set aside a conviction of a Mi'kmaq man, Joshua Bernard, for harvesting timber on Crown land, ruling that Bernard had an unextinguished treaty right to harvest and sell timber.[8] Some areas have witnessed agreements that provide for co-management of forest resources, such as the agreement between Quebec and the James Bay Cree, and the agreement between the government of Newfoundland and Labrador and the Innu (Ross and Smith 2002; Wilson and Graham 2005).

Across Canada, First Nations have come to play an increasingly important role in the forest sector. Courts' decisions have forced governments to find ways to involve First Nations more directly in decision making, and to distribute more of the economic benefits of forestry to First Nations. Many forest companies have developed their own Aboriginal policies on consultation and accommodation and have experimented with joint ventures and other economic arrangements with Aboriginal groups.

The Rise of Multi-Stakeholderism

The increasing importance of actors beyond the traditional nexus of government and business is one of the hallmarks of the widespread trend toward new governance structures (Howlett and Rayner 2006a). As a result of the increased power and legitimacy of these players, governments have been forced to reconsider the process of policy development and implementation. One of the most prominent manifestations of this trend is multi-stakeholder consultations, where governments gather together relevant interest groups in a process designed to develop agreement on policy changes (Lindquist and Wellstead 2001). There are a number of examples of these new forums. One prominent early example is the round tables focused on integrated environment and economy in the late 1980s (Howlett 1990). Although many of these have been terminated, the National Round Table on the Environment and the Economy continues to play a prominent role and recently released a major study on the boreal forest. Multi-stakeholder processes were also used to develop the national forest strategies that first articulated the sustainable forest management paradigm in Canada (Rayner and Howlett 2007).

Multi-stakeholder consultations have been most influential in land use decision making. British Columbia has developed several comprehensive "land and resource management plans" through consensus-based exercises involving a wide range of resource-related interests

(J. Wilson 1998). Most of the planning tables were able to reach consensus, and even those that did not were quite influential on the ultimate decisions adopted by government (Thielmann and Tollefson 2009). Alberta and Ontario have also relied on multi-stakeholder consultations to develop comprehensive land use plans (Cartwright 2003). In areas such as land use planning, these consultations have become necessary in order for governments to garner legitimacy for their decisions. As a result, multi-stakeholder consultations have empowered a wide variety of policy actors beyond the traditional nexus of business and government to participate in policy decisions in a meaningful way, while at the same time greatly complicating the policy making process.

Addressing Sustainable Forest Values through Regulation and Increasing Attenuation of Tenure Property Rights

The environmental movement and assertion of a greater First Nations role in forest policy and management are the most prominent examples of a societal shift toward increased concern for the environmental and social aspects of the sustainable forestry management equation. At the provincial level, these new interests have been accommodated largely through a combination of changes in planning processes and new regulations on land use and forest practices. All provincial jurisdictions have some requirements for strategic and operational planning (as described for five provinces in Chapter 4). These have been modified to give the public and First Nations greater opportunity for participation. In addition, all the jurisdictions have substantially updated and formalized the manner in which environmental values are addressed in the forest.

Table 1.2 provides an overview of some major changes in statutes, regulations, and land use planning processes in British Columbia, Alberta, Ontario, Quebec, and New Brunswick. There was a major burst of new legislation in the early to mid-1990s, as typified by BC's Forest Practices Code, Ontario's Crown Forest Sustainability Act, and Alberta's Ground Rules.[9] More recently, there has been a new wave of reform to modify regulatory frameworks to incorporate new understanding of various aspects of sustainable forest management, especially the protection of biodiversity and the incorporation of frameworks for criteria

Table 1.2 Major forest sector planning and regulatory changes in five provinces

Year	Province	Item
1982	New Brunswick	Crown Lands and Forest Act
1986	Quebec	Forest Act
1988	Ontario	*Timber Management Guidelines for Protection of Fish Habitat*
1988	Quebec	Regulation respecting standards of forest management for forests in the public domain (RSFM)
1989	New Brunswick	Clean Water Act
1991	Ontario	Code of Practice for Timber Management Operations in Riparian Areas
1991	Quebec	Draft strategy on forest protection, followed by public consultations on the strategy
1992	British Columbia	Protected Areas Strategy launches to double protected areas to 12 percent
1993	Alberta	Alberta Forest Conservation Strategy begins
1993	Ontario	Policy Framework for Sustainable Forests approved by cabinet
1993	Quebec	Forest Act updated
1993	Quebec	RSFM updated
1994	British Columbia	Forest Practices Code Act
1994	Alberta	Alberta Timber Harvest Planning and Operating Ground Rules
1994	Ontario	Crown Forest Sustainability Act
1994	Ontario	Class Environmental Assessment of Timber Management on Crown Lands in Ontario
1994	Quebec	Forest Protection Strategy
1994	New Brunswick	*Forest Management Manual for Crown Land*
1995	Alberta	Special Places program begins
1995	Ontario	*Forest Operations and Silviculture Manual*
1995	Ontario	*Scaling Manual*
1995	Ontario	Conservation Strategy for Old-Growth Red and White Pine Forest Ecosystems
1996	Alberta	Forests Act updated
1996	Ontario	*Forest Management Planning Manual*
1996	Quebec	Forest Act updated
1996	Quebec	RSFM updated
1996	Quebec	Biodiversity Strategy and Action Plan

▶

◄   Table 1.2

Year	Province	Item
1996	Quebec	Review of forest biodiversity
1996	New Brunswick	*Watercourse Buffer Zone Guidelines for Crown Land Forestry Activities*
1997	British Columbia	Forest Practices Code Act streamlined
1997	Alberta	Alberta Forest Conservation Strategy concludes
1997	Ontario	Lands for Life consultation process begins
1998	Alberta	*Alberta Forest Legacy*
1998	Alberta	*Interim Forest Management Planning Manual*
1998	Quebec	Public consultations on government review of forest system
1999	Ontario	Ontario Forest Accord and Approved Land Use Strategy
1999	New Brunswick	*A Vision for New Brunswick Forests: Goals and Objectives for Crown Land Management*
1999	New Brunswick	*Watercourse Buffer Zone Guidelines for Crown Land Forestry Activities* revised
2000	Ontario	*Forest Operations and Silviculture Manual* amended
2000	Ontario	*Scaling Manual* updated
2000	New Brunswick	*A Vision for New Brunswick Forests: Goals and Objectives for Crown Land Management* revised
2001	Alberta	End of Special Places program
2001	Ontario	*Forest Information Manual*
2001	Ontario	*Forest Management Guide for Natural Disturbance Pattern Emulation*
2001	Quebec	Forest Act updated
2002	Ontario	*Room to Grow* released
2003	Ontario	*Room to Grow* framework legislated
2003	Ontario	Crown Forest Sustainability Act updated
2003	Ontario	*Old Growth Policy for Ontario's Crown Forests*
2004	British Columbia	Forest and Range Practices Act comes into force, more results-oriented
2004	Ontario	New *Forest Management Planning Manual*
2004	Quebec	Coulombe Commission
2004	New Brunswick	*Forest Management Manual for Crown Land* (interim)

and indicators. The years 2004 and 2005 saw the Coulombe Commission in Quebec, leading to a substantial reduction in the allowable annual cut and the adoption of an ecosystem-based management framework; the implementation in British Columbia of the Forest and Range Practices Act, designed to be a more results-oriented, flexible regime; and new forest planning manuals in New Brunswick, Ontario, and Alberta.[10]

Consequently, in response to emerging societal concerns reflected in the sustainable forest management paradigm, governments have imposed new, complex, and costly conditions on Crown forest tenure holders that they must meet in order to exercise their rights. In the terms of the conceptual framework adopted for this study (see Chapter 2), more elaborate systems of planning and increasingly comprehensive and stringent environmental regulations constitute an attenuation of the property rights embedded in forest tenures.

Increasing Enthusiasm among Policy Analysts for Market and Incentive-Based Instruments as Alternatives to "Command and Control"

Although governments across Canada have been instituting an increasing number of regulatory requirements within many sectors, greater interest in alternative policy instruments has emerged among policy analysts and some governments. This trend, which has its roots in the late 1960s, has been critical of the rise of regulatory policy instruments that have become known pejoratively as "command and control," and promotes alternative policy instruments that rely on fiscal and market incentives rather than coercion. Interest in such alternative instruments increased in the 1980s as the OECD began promoting their use, and more so in the 1990s as interest in voluntary approaches further increased (Harrison 2001).

Alternatives to command and control regulations embody a wide range of instruments, with tremendous variation in potential consequences. For example, many economists advocate effluent taxes and, more recently, marketable permits. The United States has developed a thriving market in tradable permits for sulphur dioxide, and a number of jurisdictions are embarking on systems of marketable permits (or "cap and trade") in an effort to reduce greenhouse gas emissions (Stavins 2007). These instruments allow firms a great deal more flexibility in their choice of strategies to achieve required standards than do command and control, but still have a regulatory basis. There are also approaches that have little, if any, regulatory basis, such as voluntary instruments that support sustainability reporting (KPMG Global Sustainability Services 2005).

In Canadian forestry, there has been a significant amount of criticism of over-reliance on command and control regulation (e.g., Pearse 1998; Stanbury and Vertinsky 1998) but relatively little innovation by governments in designing new policy instruments. Marketable permits have yet to be introduced in the forest sector.[11]

Consistent with the emergence of the "new governance," there has been greater attention to so-called results- or performance-based regulations. The emergence of criteria and indicators as a dominant framework for forest management has encouraged policy makers and forest managers to focus on performance and results. This trend is particularly evident in British Columbia, where the Forest Practices Code has been replaced by the more results-based Forest and Range Practices Act. As we will see in Chapter 4, this approach can still be considered "command and control," but it does provide operators with greater flexibility in meeting government-specified objectives.

One of the biggest changes in policy instruments employed in forestry has been the marked shift toward reliance on voluntary instruments. As part of a larger society-wide trend toward corporate social responsibility, numerous forest companies have undertaken sustainability reporting. Most important, voluntary certification to independent standards for sustainable forest management has become a fundamental part of forest sector governance.

Rise of Certification

Perhaps the most dramatic trend in governance in the forest sector is the rise of private certification organizations (Cashore, Auld, and Newsom 2004). In 1993, frustrated with the lack of progress on a global forestry convention, environmentalists formed the Forest Stewardship Council

(FSC). The organization developed a set of international "principles and criteria" for sustainable forestry, and accredited various certification organizations to audit and certify that companies met the standards. Regional standards specific to areas around the globe were also developed. FSC Canada has been working on the development of regional standards in the Maritimes, the Great Lakes-St. Lawrence region, British Columbia, and the boreal forest zone. The BC standard was awarded preliminary accreditation in 2003, and formal accreditation in 2005 (Tollefson, Gale, and Haley 2009). In August 2004, the boreal standard was accredited by FSC International. This standard is applicable to over 75 percent of Canada's forestland. It is anticipated that the adoption of the boreal standard will accelerate the FSC's certification of Canadian forestlands.

The FSC has had a much bigger impact on forest management than the amount of certified area suggests (McDermott and Hoberg 2003). In an effort to respond to the emergence of the FSC, the forest industry and governments across Canada, building on the CCFM's sustainable forest management framework, collaborated in the development of a sustainable forest management standard under the Canadian Standards Association. In 2002, the Forest Products Association of Canada (FPAC) announced it would require all of its members to be certified by independent organizations by 2006.[12] In its 2005 annual report, the FPAC claimed that it was "95% of the way there" (Forest Products Association of Canada 2006, 25). In the United States, the forest industry developed its own Sustainable Forestry Initiative (SFI) certification program. Because so many Canadian forest products are sold into the US market, a number of Canadian firms have chosen to get certified by the SFI standard. Figure 1.1 shows the number of hectares of forestland certified by each of these three schemes.

Certification is an important trend in forest governance for several reasons. First, private certification organizations, and the consumer preferences that they purport to represent, have the potential to influence forest management decisions beyond those required by government. As a result, forest firms face an increasingly complex and challenging rule environment. Second, the political struggle over policy making has shifted to the market arena, where competing certification organizations

Figure 1.1   Forest certification in Canada by program (mid-2009).

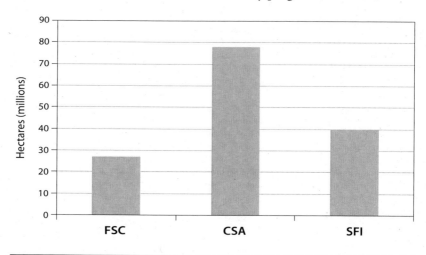

Source: Canadian Sustainable Forestry Certification Coalition, http://www.
certificationcanada.org/english/status_intentions/status.php

are battling for legitimacy in the marketplace and interest groups are taking their arguments to the governing bodies of the new certification organizations.[13] Finally, governments are struggling to redefine their own role, to take into account the emergence of the certification phenomenon. Government policies are still in place and, in fact, play a fundamental role in all certification schemes in that all schemes require compliance with applicable laws. But governments now face conflicting pressures. On the one hand, they are tempted to economize on administrative resources by relying on certification as effective compliance with sustainable forest management standards. On the other hand, they are reluctant to abandon their role in being the primary instrument for the protection of public values in the forest.

Rethinking the Level of Government Jurisdiction

The discussion thus far has focused on trends in the diffusion of governmental authority to a broader network of actors, including nongovernmental organizations and First Nations, as well as pressures to change the nature of policy instruments on which governments have relied. Additional pressures have occurred to change the level at which forest policy has historically been made. In the forest sector worldwide, there have been strong pressures toward both internationalization and decentralization (Gluck et al. 2005). But in Canada, changes in this area have been less significant than in other jurisdictions.

Canadian forest policy has been dominated by provincial governments (Howlett 2001b). At various times in recent years, the Canadian government has sought to increase its role, and some important trends have increased its leverage. Given its extraordinary high export dependence, the federal government has always had leverage through its trade powers, and this has been significant as the softwood lumber dispute with the United States has persisted. The federal government also has significant potential to assert jurisdiction over forest management through its powers over First Nations, fisheries, migratory species, and species at risk. The federal government's commitment to international treaties may also increase its desire to spur provincial action. Thus far, however, the federal government has declined to mount a significant challenge to provincial jurisdiction in forest policy.[14]

There are indications of an increasing role for "national," as opposed to federal, organizations, as indicated by the development of national forest strategies and the related sustainable forest management framework through the auspices of the Canadian Council of Forest Ministers and other organizations (Rayner and Howlett 2007). As discussed above, these frameworks have been influential in how provincial governments have organized and reported on their forest management policies, but there is little evidence that they have had any substantial impact on provincial policies. The primary reason for this limited impact is that the criteria and indicators approach identifies the values that should be addressed and ways to measure progress toward them, but it does not, at least in the Canadian version, contain specific standards or settings. For example, the CCFM framework says that you should protect ecosystem diversity, and tells you how to measure it, but it does not contain specific directions about how, or how much of, ecosystems should be protected. Those decisions are deferred to the provinces with legal jurisdiction. As a result, the criteria and indicators sustainable forest management framework has become important in organizing and communicating provincial policies, but it has not been very influential on the substance of those policies (Howlett, Rayner, and Tollefson 2009b).

Like many other areas of public policy, forest policy in Canada has experienced significant pressures toward internationalization (Bernstein and Cashore 2000; Gluck et al. 2005). The section on the environmental movement described how environmentalists have succeeded in mobilizing international market pressures, and the certification movement is part of that same trend. Several multilateral agreements, particularly the Convention on Biological Diversity and the Kyoto Protocol, have indirect implications for forestry. Efforts to create a binding legal instrument through some type of forest convention have not succeeded to date, and have defaulted to "soft law" encouraging nations to adopt national programs and criteria and indicators for sustainable forest management (Gluck et al. 2005). The increasingly widespread use of the sustainable forest management framework of criteria and indicators suggests some convergence on the conceptualization of and methodology for forest management. A coordinated framework and language helps in "explaining the direction of Canadian forest policies to the world" (Rayner and Howlett 2007, 656). But internationalization of the sustainable forest management paradigm does not yet reflect strong international influence on the substance of forest policy in Canada.

One of the most important trends in forest governance worldwide is decentralization of control over forests from national governments to local, frequently indigenous, communities (Larson and Ribot 2004; Gluck et al. 2005). In Canada, the National Forest Strategy contains a commitment to increase the extent of community forestry:

2.1 Develop and adapt forest legislation and policies to provide involvement of forest-based communities in sustainable forest management decision making and implementation.

2.2 Expand the area and use of community-based tenure systems and resource allocation models in remote, rural regions of Canada to increase benefits to Aboriginal Peoples and forest-based communities.

However, despite the increased rhetoric supporting community forestry, and substantial political activity in this area in British Columbia over the past several years (Haley 2002; McCarthy 2006), it has yet to have a significant impact on governance. The total amount of allowable cut allocated to community forests nationwide is still considerably less than 1 percent.

Forest policy in Canada will continue to be buffeted by a complex mixture of pressures toward internationalization and devolution. But unlike the other areas discussed in this chapter, what is remarkable about the vertical division of power in Canadian forest policy is not change but stability in the extent to which provincial governments dominate forest tenure policies.

Conclusion

The past several decades have seen the simultaneous emergence of sustainable forest management as the dominant policy paradigm and dramatic changes in the governance of the forest sector. The business-government partnership that dominated the sustained yield era has given way to a multipartite, multi-stakeholder system involving a broader range of actors representing a broader mix of values, and a diffusion of power. First Nations have emerged to challenge the very control by the Crown of forest resources, a development that has tremendous implications for the allocation and administration of forest tenures. Pressures from environmental groups that have resulted in a multiplicity of forest planning and practice requirements have led to much greater attenuation of Crown forest tenure holders' property rights. At the same time, certification organizations have emerged to form a type of private government where non-governmental organizations are creating rules that are influencing private management decisions.

Given the uncertainties in the direction of certification and especially Aboriginal issues, there does not seem to be a new equilibrium yet in the sense of a new, stable, well-understood relationship between government, the forest industry, and other actors in the forest policy arena. It is in the context of these broader changes and uncertainties that forest tenure policies are seeking to pursue the goals of sustainable forest management.

A General Framework for a Comparative Analysis of Canadian Crown Forest Tenures

To describe Canada's complex Crown forest policies in a way that allows interprovincial comparisons to be made, and their impacts to be critically examined, requires appropriate analytical frameworks. Although a number of analytical frameworks could be used in pursuit of this objective, in this book, a property rights framework is adopted as the principal analytical tool. We believe this approach to be appropriate because variants in Crown forest tenures may be characterized as variations in property rights, the implications of which may be used to assess the economic, ecological, and social goals of sustainable forest management. The purpose of this chapter is to describe some of the concepts underlying this methodology and introduce the property rights approach to the comparative analysis of Crown forest tenures.

Before developing the analytical framework, it is important to establish some useful definitions. The terms "institution" and "property right" are used to convey a wide variety of concepts in both scholarly and lay usage. In the next section, we begin with a brief overview of the literature on these expressions and then provide working definitions of the terminology that has been adopted for the purposes of this study. The importance of property rights in market economies is then discussed and why they comprise important instruments of public policy is explored.

What Is an Institution?

This book is concerned with institutions and institutional change. However, perceptions of what comprises an institution and how it functions vary considerably both within and across disciplines.

A small sample of the ways in which the term "institution" has been defined is presented in Table 2.1. An examination of this table reveals key differences, but also several similarities, among scholarly definitions of institutions. One of the key differences relates to the differentiation of "organizations" and "institutions." Some economists stress that the distinction between institutions and organizations is needed in order to differentiate between rules of the game and players in the game, but some sociologists and political scientists incorporate organizations into their definitions of institutions. On the other hand, rules are a common element of all definitions. However, there are different kinds of rules. Elinor Ostrom and her colleagues, in their institutional analysis and development (IAD) framework (Ostrom 1990), distinguish three tiers of rules:

- operational rules that comprise decisions about when, where, and how to do something;
- collective choice rules that determine how operational rules can be changed and who can participate in this type of decision; and
- constitutional choice rules that establish how collective choice rules are made.

Collective and constitutional choice rules may be viewed as rules of policy process that govern what is allowed during the specification of operational rules. Operational rules, which specify what may and may not be done, are formulated within a framework of process-based rules. For example, constitutional choice and collective choice rules may dictate that public consultations must take place as integral components of processes designed to prepare operational rules, and even specify the forms such consultations might take.

Within the context of these three types of rules, some authors (e.g., Hoberg 2001b) consider institutions to apply only to the process-oriented rules (i.e., collective choice and constitutional choice rules), whereas other authors (e.g., Ostrom 1999) use the term "institutions" to refer to all three types of rules.

Whereas process-oriented rules have an impact only on those organizations involved in policy making, operational rules may affect a broad spectrum of society. For example, governments, NGOs, and Aboriginal groups may all be involved in developing processes for setting aside and establishing operational rules for protected areas and must work within the constitutional choice rules that have been developed to regulate this policy-making process. However, the operational rules that are put in place dictate what all users of protected areas can and cannot do.

For purposes of this study, the following definitions of the terms "organization" and "institution" have been adopted: an organization refers to a group of people sufficiently motivated toward one or more common purposes such that they agree to obey the rules they create; institutions are rules that organizations make, including process rules (i.e., collective and constitutional choice rules) and operational rules.

What Is a Property Right?

Luckert (2005) has suggested that property rights arise within the context of operational rules devised within a framework of constitutional choice and collective rules. As such, property rights arise from frameworks of operational rules under which most of Canada's forestlands are administered. Consequently, a clear understanding of the nature of property rights is essential to any exploration of Canadian forest policy.

Definitions of "property rights" have evolved as generations of scholars from various disciplines have attempted to understand their institutional roles in determining how societies are organized and function. In the following discussion, we begin with definitions that establish broader concepts of property rights before focusing on a definition that we use for our analyses. A collection of "property rights" definitions from the economics literature is shown in Table 2.2.

People often speak of "owning an asset" – a house, a vehicle, or a forest. However, following the definitions in Table 2.2, it is evident that the term "property right" implies a much more nuanced concept than is implied in this popular notion of ownership. In fact, it is rights, never objects, that are owned. A property right permits its holder to use an

Table 2.1 How different disciplines define institutions

Discipline	Definition
Economics	"Institutions are the rules of the game of a society or more formally are the humanly-devised constraints that structure human interaction. They are composed of formal rules (statute law, common law, regulations), informal constraints (conventions, norms of behavior, and self-imposed codes of conduct), and the enforcement characteristics of both. Organizations are the players: groups of individuals bound by a common purpose to achieve objectives." (North 1993, 5-6)
Sociology	"Institution and institutionalization are core concepts of general sociology. Across the social sciences, scholars reach for these terms to connote, in one fashion or another, the presence of authoritative rules or binding organization ... This usage conforms to what may be the core denotation of institution in general sociology, that is, an institution as an organized, established, procedure. These special procedures are often represented as the constituent rules of society (the 'rules of the game') ... Some scholars invoke *institution* simply to refer to particularly large or important associations. Others seem to identify institutions with environmental effects. And some simply equate the term with 'cultural' effects, or with historical ones." (Jepperson 1991, 143, emphasis in original)
Political science	"I will use the term institution ... to refer to the shared concepts used by humans in repetitive situations organized by rules, norms, and strategies (Crawford and Ostrom 1995). By rules, I mean shared prescriptions (must, must not, or may) that are mutually understood and predictably enforced in particular situations by agents responsible for monitoring conduct and for imposing sanctions. By norms, I mean shared prescriptions that tend to be enforced by the participants themselves through internally and externally imposed costs and inducement. By strategies, I mean the regularized plans that individuals make within the structure of incentives produced by rules, norms, and expectations

▶

◄ Table 2.1

Discipline	Definition
	of the likely behavior of others in a situation effected by relevant physical and material conditions." (Ostrom 1999, 37)
	"*Operational rules* directly affect the day-to-day decisions made by appropriators concerning when, where, and how, what information must be exchanged or withheld, and what rewards or sanctions will be assigned to different combinations of actions and outcomes. *Collective-choice rules* indirectly affect operations choices. These are rules that are used by appropriators, their officials, or external authorities in making policies – the operational rules – about how a CPR should be managed. *Constitutional-choice rules* affect operational activities and results through their effects in determining who is eligible and determining the specific rules to be used in crafting the set of collective-choice rules that in turn affect the set of operational rules." (Ostrom 1990, 52, emphasis in original)
	"The term institution refers to many different types of entities, including both organizations and the rules used to structure patterns of interaction within and across organizations." (Ostrom 1999, 36)
	"Institutions are rules and procedures that allocate authority over policy and structure relations among various actors in the policy process." (Hoberg 2001b, 11)
History	"How do historical institutionalists define institutions? By and large, they define them as the formal or informal procedures, routines, norms and conventions embedded in the organizational structure of the polity or political economy. They can range from the rules of a constitutional order or the standard operating procedures of a bureaucracy to the conventions governing trade union behavior or back-firm relations. In general, historical institutionalists associate institutions with organizations and the rules or conventions promulgated by formal organization." (Hall and Taylor 1996, 938)

Table 2.2   Some economists' definitions of property rights

Source	Definition
Fischer (1923, 27)	"The liberty or permit to enjoy the benefits of wealth while assuming the costs which these benefits entail."
Dales (1968, v)	"The set of social rules that on the one hand give individuals the right to use their 'property' in certain ways and on the other hand forbids them to use it in other ways."
Ciriacy-Wantrup and Bishop (1975, 714)	"[The] bundle of rights in the use and transfer (through selling, leasing, inheritance, etc.) of natural resources."
Dahlman (1980, 70)	"The rights of decision making with regards to an asset that is recognized within any given social setting."
Bromley (1991, 15)	"[The] capacity to call upon the collective to stand behind one's claim to a benefit stream."

asset in order to enjoy the resulting stream of benefits subject to certain conditions, obligations, and prohibitions. Drawing on the extensive literature on the theory of property rights, including the definitions presented in Table 2.2, we can define a property right as a socially sanctioned and enforceable claim of an individual or group to a stream of benefits resulting from the use of an asset subject to the restrictions and conditions society places on the use of the asset concerned.

Following this definition, a property right can be described in the following terms:

- a description of the asset concerned and the benefit stream to which the right is granted;
- the relationship between the property right holder and society at large, as defined by the prohibitions and conditions on the use of the right (often referred to as attenuations of the right);

- the extent to which the property right is recognized and enforceable; and
- recognition of the claimant(s) to the right to the exclusion of other members of society.

Below, we examine each of these qualities in more detail.

Assets and Benefit Streams

An asset, which was defined by Dales (1968) as "a bundle of potential utility-yielding services that can be used in alternative ways," can give rise to many different kinds of benefits, each with the potential of being the subject of a property right. For example, in the case of forests, separate property rights might be established to commercial timber of different types and species, botanical food products, medicinal products, forage and shelter for livestock, wildlife, various recreational services, water, and subsurface minerals. Property rights to these varied products may be held as a single property rights bundle. Alternatively, the right to use one or a combination of the products may be held by different individuals or groups. In Canada, rights to the many different forest products frequently have separate owners. For example, although governments generally retain rights to the productive capacity of the land itself; water associated with forests, fish, and wildlife; and many types of recreational use, various licensing arrangements create rights to trees for industrial use, subsurface minerals, fur-bearing animals, and certain recreational services, such as guiding, ecotourism, and skiing. In some provinces, for example Alberta and Quebec, separate property rights are often granted to different licensees to coniferous trees and broad-leafed trees within the same forest.

Prohibitions and Conditions

Property rights are rarely, if ever, completely unfettered. Assets can be enjoyed only within the context of conditions that society places on their use. Consequently, the ways in which property holders can exercise their rights are circumscribed by law and/or by custom. These restrictions, or attenuations, may take many forms, from outright prohibitions on certain uses to regulations controlling the way, or degree to which, a right can be exercised. Governments frequently retain a financial interest in the benefits that are generated when a right is exercised. This interest is usually asserted in the form of taxes or other charges, such as stumpage fees in the case of trees, levied against the property right holder's benefit stream.

Because the use of forests frequently has far-reaching public consequences – environmental, economic, and social – the property rights attached to them are often highly attenuated. Restrictions may include where, how, and how much timber must be harvested; how the land is treated after harvesting; how tree crops are tended and where; and sometimes to whom timber can be sold. However, public policies may also enhance the benefits of right holders. For example, some forest licensing arrangements provide for payments to licensees to defray the costs of reforestation or protection of the forest from fire and disease.

The value of a property right is determined by the magnitude of the benefits to which it provides access and the cost of meeting the conditions under which it can be exercised. All attenuations – whether they are levies that directly reduce benefits or requirements that increase operational or managerial costs, delimit transfers, or restrict markets – reduce the net value of a right. Conversely, all arrangements that augment the benefits that property rights holders receive increase the value of their rights.

Recognition and Enforcement

Property rights, as important components of a society's institutional framework, must have the general support of society and be defined, awarded, and upheld through society's laws. Unless society recognizes a property right and is willing to defend and enforce it, it is unlikely to endure. For example, many of Canada's Aboriginal people have claimed title to their traditional territories for decades. However, such claims met with little success until they were generally recognized by the courts as having legitimacy within Canadian law.

A property right, as a legal claim to wealth resulting from the use of an asset, is of value only if it is enforceable. That is, if those other than the right holder(s) can be excluded from enjoying the benefits to which the right is granted. Legal systems, both criminal and civil, are to a very large extent concerned with the enforcement of property rights. Nevertheless, legal protection of a right does not guarantee the protection of benefits for the property holder; if protection of a property right is difficult and therefore costly to implement, it may not be enforced. This is often the case with products of forests. In some places, the Russian Far East and Indonesia for example, illegal harvesting of timber is rife and sometimes exceeds the harvests of those who hold legal timber-harvesting rights (Glastra 1999; Ravenel, Granoff, and Magee 2004). In other cases, difficulties surrounding the enforcement of a right may prevent clear rights from being created in the first place. For example, in British Columbia, harvesting of valuable crops such as mushrooms and decorative greenery is difficult and costly to control, and benefits may be received by anyone who wishes to harvest them, although legally the rights in most cases are held by the Crown.

Whether a right is enforced depends, to some extent, on the value of the right being protected relative to enforcement costs.

Claimants to Rights

Bromley (1991) placed those who hold property rights into four categories or, as he referred to them, regimes:

- private property
- state or public property
- common property
- open-access property.

Individuals, or legal entities such as corporations, cooperatives, societies, or partnerships, may hold private rights. In the case of public property, rights are vested in the state and controlled and administered by governments – the situation for 93 percent of forestland in Canada (Canadian Forest Service 2005). Common property rights are held jointly by a defined group – for example, a community, tribe, or clan – to the exclusion of all others in society. Rules under which these rights can be exercised are determined and administered by the group and are frequently a matter of custom and tradition. First Nations in Canada typically hold rights to forests and other natural resources in common. Such arrangements have attracted a great deal of scholarly interest in recent years.[1] If no property rights to an asset exist, everyone has an equal right to the benefits it generates, and a state of "open-access" prevails. Open-access, then, refers to an absence of rights – "everyone's property is nobody's property."

What Is a Forest Tenure?

Following from the discussions above, it is evident that forest tenures are means by which governments grant benefit streams from forest resources to individuals or organizations, subject to numerous operational rules that are conditions of holding tenure. As such, we define forest tenures as property rights to forest resources granted to private firms by governments.

In Canada, Crown forest tenures that delineate property rights to the nation's public forests influence the behaviour of both public and private agents in the forest sector and, consequently, have been key instruments of public forest policy since the earliest years of colonization.

Conceptual Linkages among Organizations, Institutions, Markets, Property Rights, and Forest Tenures

Table 2.3 presents a summary of the definitions adopted for this study. In Figure 2.1, the relationships between organizations, institutions, and property rights are portrayed schematically.

At the top of Figure 2.1 are three boxes representing organizations, institutions, and markets for goods and services. Organizations, including governments; lobby groups, such as private firms; NGOs; and, in Canada, First Nations, create institutions that comprise a set of rules that govern the behaviour of the individuals and groups within society and how they interact one with another. Government is highlighted as being a higher order organization than others in that it has more authority to establish binding rules and ensure that they are enforced. Rules may determine policy-making processes in the case of constitutional

Table 2.3 Definitions adopted in this study

Term	*Definition*
Organization	A group of people sufficiently motivated toward one or more common purposes such that they agree to, and obey, rules that they create.
Institutions	Rules that organizations make, including process rules (i.e., collective and constitutional choice rules) and operational rules.
Property right	A socially sanctioned and enforceable claim of an individual or group to a stream of benefits resulting from the use of an asset subject to the restrictions and conditions society places on the use of the asset concerned.
Forest tenure	Property rights to forest resources granted to individuals or organizations by governments.

Figure 2.1 Conceptual linkages among organizations, institutions, markets, property rights, and forest tenures

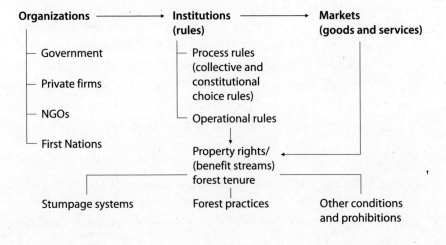

and collective choice rules, or policy prescriptions in the case of operational rules.

Institutions create the policy environment within which market goods and services are produced and exchanged. The benefit streams that result from market transactions, within the context of operational rules, define property rights that govern the behaviour of property rights holders and determine how effectively property rights function within a market economy to achieve social objectives.

In the case of the Canadian forest sector, these property rights are embodied in forest tenures. Operational rules are predominantly implemented through tenure arrangements that provide their holders with rights to use forests in certain ways and specify the conditions and prohibitions that must be observed in order to exercise these rights. Prominent among these conditions are the rules governing the payment of stumpages and other levies to governments and how forests should be managed – that is, forest practices.

Roles of Governments and Markets

It is apparent from the above discussion that how economic systems behave in the pursuit of public goals is largely a function of how property rights are specified. Exchanges of goods are not merely trades of physical assets. They are trades in benefit streams generated by assets within the context of a set of societal rules, which ideally ensure that market behaviour serves the public purpose. These rules help determine the market value of each benefit stream generated by an asset and, hence, the value of the asset itself. For example, in most Canadian provinces, logs harvested under licence from public land must be processed by domestic mills unless an export permit, which is frequently denied, is obtained. These market restrictions may significantly reduce the market value of logs and, consequently, the value of the benefit stream the owner of a harvesting licence can expect to generate. When such a licence is sold, its market value reflects the various requirements, including the virtual prohibition placed on log exports, that must be met in order to exercise the rights granted.

Given the importance of property rights in the determination of market values and the behaviour of their holders, it follows that governments can use the specification of property rights as an important instrument in the pursuit of social goals. In some cases, governments may choose to entrench virtually unfettered property rights and allow free markets to allocate resources and distribute economic benefits. In other cases, such a strategy may have outcomes that are considered to be contrary to the public good. When this occurs, markets are said to have "failed," and corrective action in the form of regulations that often take the shape of modifications to property rights arrangements, are introduced. However, devising "optimum" regulations that correct market failures in an effective and efficient manner and result in improved social welfare is a difficult task. Governmental regulation of market processes often have unforeseen effects and outcomes that are contrary to their intent. In such cases, "government failures" are said to have occurred. Sources of market failures in the forestry sector are discussed in depth by Boyd and Hyde (1989); Wolf (1988) provides a comprehensive discussion of government failures.

In an ideal world, where private interests coincide with public objectives, market forces could be given full reign to regulate the production of goods and services and the distribution of wealth. Where public and private goals fail to coincide, public regulation may occur to steer private behaviour toward the public interest. But identifying the best public interest in the real world is not straightforward. The perception of public interest is, to a great extent, a matter of political ideology and often reflects personal circumstances. Various groups within society tend to promote their own self-interest as being in the best public interest. For example, in the Canadian forest sector, widespread regulations require licensees to direct their logs to designated mills. Organized labour and the communities where the mills are located applaud such policies as being in the best public interest, but the forest industry may see such policies as impediments to economic efficiency and the ability to maximize the value of the benefit stream from their licence to harvest public timber. Also, public goals and perceptions of best public interest

Figure 2.2   Public vs. private control of forest property

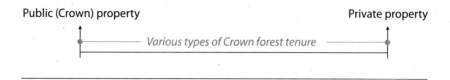

change over time. Twenty years ago, terms such as "sustainability," "biodiversity," and "ecological integrity" were barely in the lexicons of those responsible for managing Canada's forests. Today, such terms are commonly understood by a significant proportion of the population and are major driving forces of public forest policy.

As public concerns have changed, the powers of interest groups have waxed and waned, and provincial and federal governments of differing political stripes have come and gone. Against this backdrop of social and political change, forest policies have evolved that embody complex forest tenure systems. All individual tenure arrangements allow their holders to be governed by market forces to a greater or lesser extent, but all embody considerable amounts of public regulation.

Gradients of Property Rights

To understand the complexity of Crown forest tenure systems that have emerged from decades of policy evolution, it is useful to envisage different types of forest tenure as falling along a spectrum (see Figure 2.2). Governments create forest tenures by transferring from the public to the private domain certain rights to use and benefit from forests. The actual rights transferred (discussed below), the period over which they can be enjoyed, the limits imposed on these rights, and the terms and conditions under which they can be exercised and maintained determine where a Crown forest tenure lies on the public-private spectrum. Some tenures have certain rights that approach complete private ownership – the right to transfer to a third party, for example; others are severely constrained and are situated toward the public end of the spectrum.

At one extremity of the spectrum (see Figure 2.2) is public property that is characterized by complete public title to and control over

Figure 2.3   The exclusiveness of property spectrum

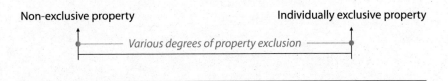

forestland and all the products it is capable of producing. At the other extremity is private property under which private individuals or groups hold unfettered title to forestland and all its products. Unfettered private control over forestland is unknown in Canada. In granting private property rights, governments generally retain control over wildlife, fish, water, and subsurface minerals, and management rights are attenuated to varying degrees. For example, private forestland owners must obtain a federal permit to export logs, and in certain provinces private holdings are subject to some form of forest practices regulations. Similarly, completely unfettered public control over forestland is also unknown. Even in cases where governments prohibit the private harvesting of timber, there are other goods and services, such as many recreational pursuits, to which private individuals enjoy rights of access.

Viewing property rights in this way provides a more continuous, rather than dichotomous, view of the private versus public property regimes. The more doctrinaire view of private versus public ownership, which can obfuscate important issues of public welfare in forest policy debates, is avoided by the notion that whether ownership is private or public is a matter of degree.

Likewise, a property-rights spectrum regarding exclusion can be used to analyze more closely common and open-access property rights regimes (see Figure 2.3). Exclusivity refers to the ability of a property right holder to exclude others from gaining access to the benefit stream that the right provides. Whether a property regime is regarded as open-access or common depends on the degree of exclusivity to the right.

At one extreme, a property right may be exclusive to an individual holder. For example, an individual wishing to occupy a campsite generally purchases a right that allows the permit holder to exclude others from using facilities within the boundaries of the site – the tent site, picnic table, and firepit for example. On the other hand, the view from the campsite may be at the other extreme: completely non-exclusive and freely available to other campers and anyone who wishes to visit the campground. Situations where exclusivity is completely absent are often referred to as "open-access." Between these two extremes there may be varying degrees of exclusivity. For example, if a group – a Boy Scout troop, perhaps – has rented a campsite, it has purchased access to a benefit stream for all its members, to the exclusion of others. When there are sufficient numbers of people seeking to coordinate their use of a property right but still able to exclude other members of the general population, rights are considered to be "common property."

Concepts of exclusivity and public and private property rights may be considered independently. That is, there may be largely exclusive public property rights, or there may be non-exclusive public property rights. For example, Canadian provincial governments, before transferring timber harvesting rights to private firms, have exclusive rights to the benefit streams the timber resource can provide. However, in most cases, the land on which the trees grow, although usually referred to as public land, is, through common usage, generally regarded as open-access, or non-exclusive, and can be used by anyone who wishes to participate in outdoor activates such as hiking and birdwatching. Similarly, private property rights may have varying degrees of exclusivity. In Sweden, for example, where about 80 percent of forestland is "privately owned," a common law – *allemansrätten* (the right of all men) – allows the public, including non-residents, rights of access to all forestland – public and private – to hike through the natural landscape, camp for one night, and gather berries, mushrooms, and wild flowers, provided there is no property damage resulting in economic loss and that the privacy of the landowner is respected (Saastamoinen 1999). Thus, significant components of the benefits produced by Sweden's private forestland are open-access.

The task of analyzing property rights is more complex than suggested by the above examples, for there are many other characteristics of property rights, as we will see in the next section, that interact with each other and, in aggregate, determine the impact of a particular right on the behaviour of its holder and the implications of this behaviour for society at large. For instance, in the above example illustrating the concept of exclusivity, the benefit stream received by a camping individual or group is dependent on the number of days purchased. Thus, duration of the benefit stream is a key defining feature of a property right.

When analyzing property rights, it is important to avoid absolutes but, instead, consider property rights as matters of degree – for example, degrees of private versus public control and degrees of exclusivity. Thus, in our analytical framework described below, we attempt to avoid definitive concepts of property rights such as public, private, or common property, and instead seek to describe rights in terms of several key features that are thought to significantly impact benefit streams, thereby having important behavioural influences and policy implications.

Property Rights Characteristics of Crown Forest Tenures
Property rights are often envisioned as bundles of attributes. These bundles can be disaggregated into their individual components, which are referred to as property rights characteristics or dimensions. Each of these characteristics may be regarded as a variable with alternative specifications. For example, the characteristic "exclusivity," discussed above, may vary between extremes of no exclusivity to completely exclusive rights. Likewise, the characteristic "transferability," described below, may vary from completely unfettered transferability through varying degrees of control over the transference of rights to a complete prohibition on transfers.

Several alternative systems for describing property rights characteristics exist. The basic structure of the system adopted for this study was first proposed by Scott and Johnson (1983) to describe property rights in general. This scheme was modified by Haley and Luckert (1990) to specifically describe Crown forest tenures in Canada and has been further amended for purposes of this study through the inclusion of some additional characteristics.

The analytical framework described below disaggregates Crown forest tenures into twelve components that provide information on the nature of the rights transferred from public to private sectors, how the rights are transferred, and how, and to what extent, these rights are attenuated. Each of these characteristics has a bearing on tenure holders' behaviour and, in aggregate, create complex systems of incentives and disincentives that have important public policy implications. Describing Crown forest tenures in this way provides a powerful analytical tool for comparing tenure systems and understanding the relationships between alternative tenure arrangements and actual or expected outcomes.

Each of the tenure characteristics adopted for this study will now be described. These characteristics are summarized in Table 2.4.

Table 2.4   Characteristics of Crown forest tenures adopted for this study

Characteristic	Description
1. Initial allocation of tenure rights	The way in which Crown forest tenures are awarded through various types of bidding or application processes.
2. Comprehensiveness	The extent to which a tenure grants rights to all the benefits flowing from an asset. The larger the number of rights granted, the more comprehensive the tenure.
3. Allotment type	Whether the rights granted are area based or volume based.
4. Size restrictions	The degree to which a tenure is restricted in size in terms of area or volume.
5. Exclusiveness	The extent to which an individual or group is able to, or allowed to, keep others from accessing benefits from property rights.
6. Transferability	Whether, and under what conditions, a tenure can be sold to a third party.
7. Export restrictions	Whether, and under what conditions, goods to which rights are granted can be sold internationally and/or interprovincially.

▶

◀ Table 2.4

Characteristic	*Description*
8. Duration and renewability	The period over which rights can be exercised and whether, and under what conditions, a tenure can be renewed or replaced with a similar agreement.
9. Fiscal obligations	The disbursements, such as stumpage fees, land rents, user fees, and other charges, that tenure holders must pay in order to exercise their rights.
10. Mill appurtenancy	Whether the wood harvested from a tenure must, in whole or in part, be delivered to a designated mill.
11. Operational requirements and controls	Operational requirements refer to the various stipulations that property holders must meet in order to exercise and maintain their rights. In the case of forest tenures, requirements can be broadly classified into management and harvesting. Operational controls are measures designed to monitor the performance of tenure holders and enforce the requirements.
12. Security, mutability, and compensation	Security refers to the confidence tenure holders have that governments will remain committed to honouring and protecting the rights granted. Mutability and compensation refer to the extent to which a tenure can be legally modified or cancelled during its term and, in the event of such action, whether and how tenure holders are compensated.

1. Initial Allocation of Tenure Rights

Initial allocation of tenure rights describes the way in which Crown forest tenures are awarded to tenure holders. Provincial governments use a variety of means to award tenure rights. These different processes may be placed into three main categories. First, a competitive tendering process may be used in which the tenure is awarded to the applicant submitting the highest bid. We refer to this as a *highest bid process*. Second, a process may be used in which bids are invited that include specified components other than price. In this case, a tenure is not necessarily granted to the bidder offering the highest price but to the applicant whose bid most closely meets the government's stated objectives. We refer to this as a *multi-attribute bidding process*. Third, a tenure may be directly awarded by the government to a chosen firm or individual with no recourse to any bidding or competitive proposal processes. We refer to this as a *non-competitive process*.

2. Comprehensiveness

Comprehensiveness describes the actual products and services to which tenure holders have rights; that is, from which they can enjoy benefit streams. The more inclusive a tenure is in terms of the number of resources to which it grants rights, the more comprehensive it is said to be. Fully comprehensive rights to forests would include the land itself, all botanical products, the soil, wildlife, water, fish, and subsurface minerals.

Comprehensiveness is not always easy to assess. For example, although forest tenures in Canada clearly grant rights to harvest trees, it may not be apparent whether rights to grow trees are included. Recall that a property right comprises a benefit stream that can be enjoyed subject to certain restrictions and requirements. In the case of growing trees, it is not always clear whether tenure holders are regenerating forests following harvesting because they perceive that they will receive benefits from the next crop of trees, or whether they are ensuring that a new crop will be produced as a condition that enables them to maintain their harvesting rights (Haley and Luckert 1995).

3. Allotment Type

Allotment of rights refers to whether the rights granted are area based or volume based. For area-based allotments, rights are granted to a specific area within well-defined geographical boundaries, and the tenure holder is largely responsible for deciding where harvests occur. For volume-based allotments, rights are granted to a certain quantity of product to be harvested within a broadly defined region without identifying, in advance, the actual locations in which the harvest will take place.

4. Size Restrictions

Size restrictions refer to limits placed on the area or quantity of product that may be granted under a particular tenure arrangement. In some cases, these limits are specified formally in legislation; however, there may also be customary size limits reflecting the size ranges within which certain types of tenure have been granted historically. Moreover, size specification may also refer to situations where tenure allocations are reserved for one sector of the industry – to promote small businesses, for example.

5. Exclusiveness

Exclusiveness refers to the right of tenure holders to prevent others from freely enjoying the benefits of the property to which they hold rights. At one extreme, property rights may be exclusive to a single individual or well-defined group, such as a corporation. At the other extreme, there may be open-access with no exclusivity. In between these two extremes, as we discussed earlier, there may be numbers of individuals or groups that are allowed access.

6. Transferability

Transferability indicates the extent to which tenure holders can sell, lease, post as collateral, or otherwise dispose of the property to which they hold rights. Closely allied to transferability is the notion of divisibility. Completely transferable and divisible property rights allow their holders to sell, partially or wholly, their rights to any party they wish. Completely restricted transferability prevents the sale of the right. In between these two extremes, transfers of rights may be allowed under certain conditions. For example, governments might deny the sale of Crown forest tenures to foreign-owned companies but allow domestic sales. Or, a transfer might be denied if it allows, in the opinion of the government, an unacceptable degree of industrial concentration in regional markets for timber or other inputs.

7. Export Restrictions

Export restrictions indicate whether certain products – generally logs and pulp chips – harvested under the terms of a tenure agreement can be freely exported out of the province and/or country, or whether such exports are restricted.

8. Duration and Renewability

Duration refers to the period, or term, over which a property right can be exercised. In some cases, property rights may be specified for short periods with no renewability provisions, whereas other rights may be granted over long terms, with or without the potential for renewal.

9. Fiscal Obligations

Fiscal obligations refer to the payments tenure holders must submit to the government. All Crown forest tenure arrangements require their holders to pay governments a variety of levies. These charges may have a significant impact on the value of rights by reducing the worth of the stream of benefits that a tenure holder can extract from the asset to which the rights are held.

Forest tenure holders in Canada are commonly required to pay land rents; renewal fees; various user charges, such as scaling fees; protection fees; and contributions to special funds that are used to support activities such as reforestation, silviculture, and research. However, the most important charges that all tenure holders in Canada face are stumpage fees. These are the direct charges levied by governments for Crown timber. They are generally charged on the volume of logs actually harvested and are expressed in dollars per cubic metre. Stumpage fees not only reduce tenure holders' net revenues but have significant impacts on their behaviour and, indeed, on the way in which the forest sector as a whole functions. The magnitude of stumpage fees and the way they are levied have implications for harvesting efficiency, the technology adopted by timber products' manufacturers in their mills, the timber processing sectors' product mix, and the way in which the wealth generated by timber resources is distributed (Luckert and Bernard 1993; Haley 2001). Furthermore, stumpage fees, particularly the ways in which they are determined, are the major irritant in the softwood lumber trade dispute between Canada and the United States that has plagued Canadian lumber manufacturers for almost a quarter of a century (Percy and Yoder 1987; K. van Kooten 2002; Zhang 2007).

Stumpage fees are such an important component of Canadian forest policy that in this study they are singled out for special treatment, and are the subject of Chapter 5.

10. Mill Appurtenancy

Mill appurtenancy refers to whether a tenure holder must own and/or operate a processing facility for the products covered by the tenure in order to exercise the rights granted. Appurtenancy requirements vary. In some cases, products harvested under the terms of the tenure must be directed to the appurtenant processing facility. In other cases, mere ownership of a processing facility by the tenure holder is sufficient.

11. Operational Requirements and Controls

Operational requirements refer to the various conditions tenure holders must meet in order to exercise and maintain their rights. Operational controls are the measures governments have put in place to monitor the performance of tenure holders and enforce the various regulations under which they operate.

Generally, the intent of operational requirements is to ensure that licensees exercise their rights in a manner that furthers and protects public interests beyond timber values. They can be broadly classified into two main groups: management and harvesting.

Management requirements are designed to ensure that forests are conserved and perpetuated, ecosystem integrity is protected, and a socially desirable mix of forest products is produced. They commonly include reforestation, silvicultural requirements, forest protection, and other forest practices. In recent years, in most Canadian provinces, the burden of operational requirements on holders of timber harvesting rights has increased dramatically as demands for a spectrum of non-timber forest products and environmental services has burgeoned and broad public concern for declining natural forest cover and the preservation of fully functional forest ecosystems has emerged. Many provinces have introduced forest practices legislation that has revolutionized the ways in which holders of timber harvesting rights operate and have sent shock waves through the forest sector as firms rationalize their operations to effectively absorb the additional costs that meeting the new regulations entails. Forest practices regulations are so important to the Canadian forest sector and are attracting so much debate that in this study they are the subject of a separate analysis, presented in Chapter 4.

Harvesting requirements are used to ensure "full" utilization of timber resources; control the volume of timber harvested annually or periodically; regulate the construction, maintenance, and deactivation of roads and skid trails; and ensure that, following harvesting, environmental conditions prevail that are commensurate with forest renewal and the protection of non-timber forest values. The regulatory measures taken to accomplish these objectives include utilization standards, the determination and control of the volume of timber that can be harvested annually – usually referred to as the allowable annual cut (AAC) – road standards, permissible logging methods, and the size and pattern of clear-cuts.[2] Some of these requirements are often components of forest practices regulations.

Operational controls include all mandatory planning and reporting procedures that must be undertaken by tenure holders, such as strategic working plans, operational plans, and annual reports; the measures adopted by governments to monitor and enforce standards; and the mechanisms for penalizing tenure holders for non-compliance. Such measures are necessary to ensure that regulations are being observed and working effectively. In many cases, they form an important part of forest practices regulations.

12. Security, Mutability, and Compensation

Security is somewhat different from those characteristics described above in that it does not refer to the legal or customary conditions defining a property right. Rather, it depends on the extent to which the holder of a right trusts that the socio-political system under which the right was granted will remain stable and that governments will remain committed to honouring and protecting the right. The confidence that property holders have in their rights depends on their past experiences and the probabilities they attach to future changes in public policy or behaviour that will adversely affect the way in which their rights can be exercised.

Although security is crucial in influencing investment incentives, it is difficult to measure objectively. Therefore, it will not be explicitly included in the comparisons of Crown forest tenures by province, which is the subject of Chapter 3. However, security can be inferred, to some degree, from the past performance of governments and the magnitude of social pressures for change. For example, First Nations land claims, particularly in British Columbia, have created a great deal of uncertainty about the future of long-standing forest tenure arrangements.

Mutability refers to the extent to which the conditions under which tenures are held can be modified during their terms. Such changes may take the form of enhanced rights, additional or modified attenuations, or outright cancellation of rights. Mutability differs from insecurity in that its scope is known in advance by tenure holders and, consequently, it can be more confidently incorporated into management decisions.

Provisions in tenure agreements for modifying or cancelling property rights may be accompanied by provisions that lay out how, and to what extent, tenure holders may be compensated should their tenures be modified in ways that have a negative impact on the value of their rights.

Comparing and Analyzing Canada's Crown Forest Tenures
Forest tenure systems in the Canadian provinces are complex. In total, excluding Prince Edward Island, there are more than forty discrete tenure types. Although these different tenure agreements have many features in common, each has unique attributes reflecting its evolution and the public objectives it is, or was, designed to achieve. Some tenures have long histories and have been modified over time to reflect changing social attitudes and goals. In some cases, new conditions and requirements have been superimposed on existing attenuations, resulting in complex arrangements that are difficult to understand in terms of the costs they impose on their holders and the incentives they create. Understanding tenure systems and the extent to which they either facilitate or constrain the achievement of current public aspirations is

essential to the formulation of effective, efficient, and equitable public forest policies that will promote the realization of the contemporary, overarching public goal of sustainable forest management.

In the next chapter we compare forest tenures in Canada in terms of ten of the twelve tenure characteristics described above. The other two characteristics – fiscal obligations, with its main component of stumpage systems, and operational requirements and controls, with its broad component of forest practices – have been singled out for special treatment given their importance and complexity. In each case, a special analytical framework has been developed to deal with the complexities of these characteristics. Chapter 4 provides a comparison of forest practices; stumpage systems are dealt with in Chapter 5.

Crown Forest Tenures in Canada   3

In the Introduction, we briefly described the evolution of forest policy in Canada, highlighting the general trends in provincial forest policies since Confederation. As shown, the public ownership of forestland is firmly entrenched as a Canadian institution, with responsibility for most public forests – 77 percent – coming under the jurisdiction of the provinces. Only the Maritime provinces, colonized at an early stage in Canada's history, have 50 percent or more of their forestland in private ownership, although there are substantial areas of private forestland in southern Ontario and Quebec.

In Canada, unlike most developed countries, public forests are not directly managed by a government department, a public agency, or a public corporation; rather, responsibility for the management of public forests is delegated to the private sector by means of long-term licensing and leasing arrangements. This approach, widely adopted by the Canadian provinces in their early years, is today the cornerstone of provincial forest policies. Under these arrangements, in return for exclusive, usufructory timber harvesting rights, licence holders contribute to Crown revenues through the payment of royalties, stumpage, and other levies and also assume varying degrees of responsibility for forest management. These arrangements have become known as "Crown forest tenures" and today account for most of the timber harvested from Crown lands in Canada. In this chapter, we compare Crown forest tenures both within and among provinces. Similarities are recognized and important variations in select provinces are identified. For example, British Columbia has numerous exceptions to national patterns of tenure, which are described below. The chapter concludes with an analysis of provincial forest tenure systems in relation to the goals of sustainable forest management: to what degree and how effectively do public forest policies, as embodied in Crown forest tenure systems, further these overarching goals?

Interprovincial Comparison of Crown Forest Tenures

Number and Relative Sizes of Crown Forest Tenure Types in the Canadian Provinces

There are close to forty forest tenure types in Canada's provinces, excluding Prince Edward Island. Table 3.1 shows the main tenure types by province and, for each type, the number issued, the volume of timber allocated, and its contribution to the provincial Crown allowable annual cut (AAC).

Each tenure has its own unique set of characteristics, but there are many similarities. Most provinces have one or two major tenure types that account for a majority of the province's AAC. With the exceptions of British Columbia and New Brunswick, all provinces have one major tenure type that accounts for 60 percent or more of the provincial AAC, and four have one major tenure type that accounts for close to, or more than, 80 percent of the AAC. These tenures are largely designed for, and usually held by, relatively large pulp and fully integrated forest products companies. In addition, all provinces have one or more medium to small tenure types accounting for a smaller percentage of the AAC and held by small, frequently non-integrated, manufacturing companies, logging companies, individuals, or communities. Such tenures often provide access to specified timber types or special forest products, such as Christmas trees, maple syrup, or fuelwood.

The distribution of the AAC shown in Table 3.1 for New Brunswick is misleading in that the province's main tenure type is the Crown timber licence. These tenures, which occupy almost 100 percent of the province's

Table 3.1 Crown forest tenure types in Canada by province, 2005

Tenure type	Number	Allocated AAC (m³)	Average AAC per tenure	Proportion of Crown AAC (%)
British Columbia				
Tree farm licence	18	12,636,000	743,294	17
Forest licence	254	34,936,980	743,340	47
Woodlot licence	816	1,486,680	1822	2
Community forest agreement	13	743,340	57,180	1
Timber licence	NA*			NA*
Timber sale licence	1,752	17,840,160	10,183	24
Other	NA*	6930	–	9
Alberta				
Forest management agreement	20	15,587,625	779,381	68
Timber quota	104	5,950,541	57,217	26
Commercial timber permit	Varies annually	1,171,314	–	5
Local timber permit	Varies annually	0	–	<1
Saskatchewan				
Forest management agreement	4	5,996,405	1,499,101	79
Timber supply licence (area based)	3	816,400	272,133	9
Timber supply licence (volume based)	12	114,665	9,555	1
Forest product permit	230	850,000	3,696	11
Manitoba				
Forest management agreement	3	3,330,160	1,110,053	80
Timber sale agreement	185	759,921	4,108	18
Timber permit	2,928	54,382	19	1
Ontario				
Sustainable forest licence	45	20,861750	463,594	65
Forest resource licence	NA*	11,233,325	–	35

▶

◄ Table 3.1

Tenure type	Number	Allocated AAC (m³)	Average AAC per tenure	Proportion of Crown AAC (%)
Quebec**				
Contrat d'approvisionnement et d'aménagement forestier (CAAF)		>42,434,600		>95
Convention d'aménagement forestier (CAF)		<2,233,400		<5
New Brunswick				
Crown timber licence	10	2,142,810	214,281	41
Crown timber sub-licence	80	3,076,645	38,458	58
Crown timber sale	5,000	NA*		<1
Nova Scotia				
Long-term licence and management agreement	2	840,000	420,000	82
Volume utilization agreement	23	178,400	7,757	17
Licence	0-4	2,000-5,000		<1
Permit	NA*	NA*	8,340	57
Newfoundland and Labrador				
Long-term timber licence	214	1,784,712		
Crown timber licence	1	NA*		
Timber sale agreement	3	NA*		
Commercial cutting permit	757	1,055,285	1,394	21
Domestic cutting permit	20,449			
Timber lease	9	NA*		
Freehold grant	15	NA*		

* Not available

** In the spring of 2009, the Government of Quebec tabled the Quebec Forestry Occupancy Act (Bill 57). Under this legislation, the current forest tenure system in Quebec will be entirely restructured. Under the proposed changes, current licences will be terminated. Licensees will receive a guaranteed timber supply amounting to 70 percent of their current holdings. The remaining 30 percent will be sold competitively to the highest bidder by a timber marketing board. The bill also calls for decentralization of forest management decision making to regional management boards comprising local elected officials and representatives of forest companies and other resource stakeholders. It is intended to bring the new act into force on 1 April 2013.

Crown forestland area, are held by only four companies and account for very close to 100 percent of the provincial AAC. However, holders of Crown timber licences are obliged to accommodate volume-based Crown timber sub-licences within their licensed boundaries. Crown timber sub-licences are volume-based licences held by about eighty companies – generally sawmill operators – and account for about 58 percent of the AAC. However, Crown timber licence holders are responsible for the administration of sub-licences, including planning, silvicultural requirements, the control of harvesting rates, and the collection of stumpage on behalf of the provincial government. Consequently, in terms of responsibility for forest planning and management, it could be said that New Brunswick has one major type of tenure that accounts for 100 percent of the AAC.

British Columbia has the most diversified and complex forest tenure system in the country. There are more tenure types in British Columbia (eleven including miscellaneous tenures) than in any other province. The number of individual licences in existence is greater than elsewhere, and their holders are more wide-ranging, including over one thousand holders of small woodlot licences, thirty-three communities with community forest agreements, and over 350 timber sale licence holders, many of them independent loggers and owners of small non-integrated mills. The BC government is committed to the diversification of the tenure system. Plans are currently being implemented to significantly increase the number of woodlot licences and community forest agreements (as of spring 2009, an additional eighteen communities were in the application process), and 8 percent of the AAC has been set aside for allocation to First Nations.

Initial Allocation of Tenure Rights

As discussed in Chapter 2, Crown forest tenures may be allocated competitively where tenures are advertised, sold by auction, and awarded to the highest bidder. Alternatively, there may be multi-attribute bidding where applications for tenures are assessed according to several factors – often including a monetary bid – and awarded to the applicant that comes closest to meeting public objectives. Finally, there may be non-competitive assignments where tenures are awarded directly by the

Table 3.2   The initial allocation of Crown forest tenures in Canada

Tenure type	*Allocation**
British Columbia	
Tree farm licence	PC
Forest licence	PC
Alberta	
Forest management agreement	NC, MC
Timber quota and licence	PC, NC
Saskatchewan	
Forest management agreement	NC
Timber supply licence	NC
Manitoba	
Forest management agreement	NC
Timber sale agreement	NC, MC
Ontario	
Sustainable forest licence	MC
Forest resource licence	MC
Quebec	
Contrat d'approvisionnement et d'aménagement forestier (CAAF)	NC
Convention d'aménagement forestier (CAF)	NC
New Brunswick	
Crown timber licence	NC
Crown timber sub-licence	NC
Nova Scotia	
Long-term licence and management agreement	NC
Forest utilization licence agreement	MC
Newfoundland and Labrador	
Long-term timber licence	No longer issued
Commercial/domestic cutting permit	NC

* PC – price competition; NC – non-competitive; MC – multiple criteria

minister or his or her delegate, with no reliance on a competitive process.

Table 3.2 shows, for each of the nine provinces included in this study, arrangements for the initial allocation of the two most important tenures as measured by their contributions to the provincial AAC. For most provinces, principal tenures are either awarded directly by the minister or delegate, usually pursuant to an order-in-council or by means of a process involving multiple criteria. In either case, tenures are frequently awarded following negotiations between the applicant(s) and the government. Only in British Columbia does legislation provide that new major tenures (forest licences and tree farm licences) can be awarded only following arm's-length price competition among applicants. Although British Columbia's Forest Act does provide for the direct award of forest licences, woodlot licences, community salvage licences, and licences to cut, these arrangements are reserved for tenures offered to First Nations respecting treaty-related or interim measures[1] or to individuals to mitigate the impact of government actions that affect their rights.[2]

For some smaller, short-term tenures, for example, timber quotas and permits in Alberta, price competition might take place. However, by and large, such tenures are awarded directly, sometimes with multiple criteria being considered. An exception, as in so many aspects of Canada's Crown forest tenure system, is British Columbia. In that province, over 20 percent of the AAC is sold by BC Timber Sales – an independent profit centre within the Ministry of Forests and Range – by means of short-term (four years or less), fully competitive timber sale licences. Applicants for a timber sale licence must offer a bonus bid over and above an upset, or reserve, price, the sale going to the highest bidder.

Comprehensiveness

In most cases, forest tenures in Canada provide only the right to harvest timber, sometimes of a specific species and/or type. In Ontario, sustainable forest licences and forest resource licences provide rights to "harvest forest resources," but in practice these are generally interpreted as timber harvesting rights. The only tenures that explicitly grant rights to non-timber forest products are community forest agreements in British Columbia, which provide for an exclusive right to harvest timber and the "right to harvest, manage and charge fees for botanical products and other prescribed products."[3]

Rather than provide for more comprehensive rights under a single tenure agreement, provinces sometimes provide overlapping rights to different tenure holders that provide access to a variety of forest products, including softwoods and hardwoods, pulpwood and sawlogs, organized recreational activities, water, fur-bearing animals, and fuelwood.

In some provinces, Alberta for example, tenure documents provide rights to "grow, manage, and harvest timber." However, in reality, no provinces provide clear rights to grow timber. To do so would require governments to grant equity in second-growth timber crops and rights in the productivity of the land itself – an entitlement that, in most cases, is denied by provincial statute (Luckert and Haley 1993). The absence of growing rights is reflected in the fact that most forest tenure agreements in Canada mandate reforestation and, in some cases, further silvicultural interventions to required standards (see Chapter 4).

Allotment Type

Table 3.3 shows the percentages of provincial Crown AACs allocated through area-based tenures. Most provinces allocate the majority of their AAC in this way. Six of the nine provinces included in this analysis allocate 80 percent or more of their AAC on an area basis and two – Ontario and Quebec – 100 percent. The exceptions are British Columbia and New Brunswick. In fact, British Columbia is the only province with a major industrial tenure – the forest licence – that is volume based. Forest licences account for almost half of British Columbia's AAC.

In the case of New Brunswick, about 58 percent of the AAC is harvested from volume-based Crown timber sub-licences. But sub-licences are managed by Crown timber licensees as integral components of their area-based licences and, thus, do not display the problems normally associated with volume-based tenures. It would not be unreasonable to assert that 100 percent of New Brunswick's AAC is allocated through area-based tenure arrangements.

Table 3.3   Allocation of AAC to area-based tenures by province

Province	AAC allocated to area-based tenures (%)
British Columbia	44*
Alberta	68
Saskatchewan	88
Manitoba	83
Ontario	100
Quebec	100
New Brunswick	41
Nova Scotia	82
Newfoundland and Labrador	80

* This amount includes short-term timber sale licences that account for 24 percent of area-based tenures.

In the case of British Columbia, only about 40 percent of the AAC is allocated to area-based tenures but, in terms of the managerial implications of volume allotments, a more realistic figure is a much lower 20 percent, which excludes area-based timber sale licences (TSLs) (see Table 3.3). Area-based TSLs are short-term tenures, not exceeding four years, that are sold competitively by BC Timber Sales. Following harvesting, they revert back to the Crown. These tenures, although area based, do not provide the benefits of integrated, long-term management planning over large areas. On the contrary, they are small, fragmented blocks spread over large areas, and often overlapping other tenure types. They present a major management challenge for the Ministry of Forests and Range in British Columbia, and their number and the area they occupy will likely increase over the next few years as more timber is sold in competitive markets.

Size Restrictions

With only minor exceptions, for example, woodlot licences in British Columbia, there are neither lower nor upper legislated limits on the size of Crown forest tenures in Canada. However, although few legal limits on size exist, there are unspecified customary limits on the size ranges of most tenure types. For example, Alberta would not issue a "small" forest management agreement, and many provinces have "small" tenures that are specifically designed for individuals and small firms. Average sizes of tenures awarded are included in Table 3.1. The largest tenures tend to be in the range of 500 to 1,500 thousand m³ per year; smaller tenures, still for commercial uses, are frequently in the range of 1 to 10 thousand m³ per year.

Exclusiveness

Rights granted by Crown forest tenures are generally exclusive, although there may be exclusive rights granted separately to various products, including different types or species of trees, from a single piece of land. A notable exception is the rights granted under community forest agreements in British Columbia. These tenure arrangements provide an *exclusive* right to Crown timber but neglect to specify the exclusivity of the right that is granted to manage, harvest, and market non-timber botanical products. This omission is, presumably, to avoid conflict with the large number of people who consider that they have a customary right to enter Crown land in order to harvest such products.

Transferability

Most large and medium-sized forest tenures in Canada are transferable with ministerial permission. Exceptions include Quebec, where neither major tenures are transferable, and British Columbia, where all tenures, with the exception of community forest agreements, are transferable, and generally divisible.[4] But in British Columbia, the minister may disallow transfers of tree farm licences, forest licences, and pulpwood agreements, if, in the opinion of the minister, the transfer "will unduly restrict competition" in standing timber and/or forest products markets.[5] Consequently, in British Columbia, although the minister may disallow the transfer of certain tenures, his or her reasons for doing so are subject to statutory limitations.

Export Restrictions

All provinces, with the exceptions of Saskatchewan, Manitoba, and Nova Scotia, regulate exports of unmanufactured timber products (logs and

Table 3.4 Export controls on non-manufactured timber products
by province

Province	Controls on log and/or chip exports?	All exports	International exports only
British Columbia	Yes	×	
Alberta	Yes	×	
Saskatchewan	No		
Manitoba	No		
Ontario	Yes		×
Quebec	Yes	×	
New Brunswick	Yes	×	
Nova Scotia	No		
Newfoundland and Labrador	Yes	×	

pulp chips). Generally, such products can be exported only with ministerial permission or by order-in-council. Some provinces control all exports, including interprovincial trade; in others, controls apply only to international trade (see Table 3.4). No province exercises an outright ban on log and/or chip exports.

Duration and Renewability

Table 3.5 summarizes the duration and renewability provisions of Crown forest tenures across Canada. Most of the larger Crown forest tenures, held by pulp or fully integrated forest products companies, have terms of twenty or twenty-five years. Exceptions include Nova Scotia's long-term licence and management agreements held by Kimberly-Clark and Stora Enso, which have terms of fifty years, and Newfoundland and Labrador's long-term timber licences, which were granted for ninety-nine years. Abitibi-Consolidated Inc. holds 23 such licences, with an area of 966,000 hectares, and Corner Brook Pulp and Paper Ltd. 191, with a total area of 1,956,000 hectares. Abitibi's licences expired in 2010 and Corner Brook's in 2037.[6]

Table 3.5 Duration and renewability provisions for Crown forest tenure types in Canada by province, 2005

Tenure type	Duration (years)	Renewal/replacement conditions	Evergreen
British Columbia			
Tree farm licence	25	Replaceable	Yes
Forest licence	<26	Majority are replaceable	Yes
Woodlot licence	<21	Some replaceable	Yes
Community forest agreement			
Long-term	25-99	Yes	Yes
Probationary	5	Yes	No
Timber licence	Varies	No	
Timber sale licence	<5	No	No
Other	NA*		Varies
Alberta			
Forest management agreement (FMA)	20 (except one of 30)	All renewable for 20 years	Most new FMAs
Timber quota	1-20	Yes	No
Commercial timber permit	1-5	Yes	No
Local timber permit	30 days-5 years	No	
Saskatchewan			
Forest management agreement	20	Yes	Yes
Timber supply licence (area based)	<10	Yes	No
Timber supply licence (volume based)			
Forest product permit		No	
Manitoba			
Forest management agreement	<20	Yes	Yes
Timber sale agreement	<20 (subject to review every 5)	No	
Timber permit	1	No	

▶

◄   Table 3.5

Tenure type	Duration (years)	Renewal/replacement conditions	Evergreen
Ontario			
Sustainable forest licence	<21	Yes	Yes
Forest resource licence	NA*		
Quebec			
Contrat d'approvisionne-ment et d'aménagement forestier (CAAF)	25	Yes	Yes
Convention d'aménagement forestier (CAF)	10	Yes	Yes
New Brunswick			
Crown timber licence	25	Yes	Yes
Crown timber sub-licence	<5	Yes	No
Crown timber sale	1	No	No
Nova Scotia			
Long-term licence and management agreement (LMA)	50	Yes	Of the two LMAs, one is evergreen and one is not
Volume utilization agreement	10	Yes	No
Licence	2	Yes (for one extra year)	No
Permit	1	Yes	No
Newfoundland and Labrador			
Long-term timber licence	Variable (usually 99)	No	
Crown timber licence	20	Yes	Yes
Timber sale agreement	<6	Yes	No
Commercial cutting permit	1	Yes	No
Domestic cutting permit	1	No	
Timber lease	99	Yes	No
Freehold grant	Indefinite		

* Not available

Intermediate and smaller tenures have terms ranging from a few months to ten years, but normally not exceeding five years. Exceptions can be found in British Columbia, where woodlot licences have twenty-year terms and community forest agreements may be granted for twenty-five to ninety-nine years.

Most major tenures with terms of twenty to twenty-five years are renewable – in many cases on an "evergreen" basis. "Evergreen" frequently means renewal every five years before the end of a term, following a performance review. Under such provisions, a twenty-year, five-year evergreen agreement would be considered for renewal every five years for an additional twenty years. If renewal is not granted, there is still fifteen years remaining in such an agreement. Five-year evergreen clauses are present in Saskatchewan's forest management agreements, Ontario's sustainable forest licences, Quebec's contrats d'approvisionnement et d'aménagement forestier (CAAFs), and New Brunswick's Crown timber licences. Some of Alberta's forest management agreements are renewable on a ten-yearly evergreen basis, whereas others are simply renewable at the end of their terms for an additional twenty-year period. Manitoba's forest management agreements are renewable for additional twenty-year terms on expiration. Nova Scotia's long-term licence and management agreement held by Kimberly-Clark is renewable for forty years when its current fifty-year term expires; a similar licence held by Stora Enso is renewable on a ten-year evergreen basis.

British Columbia is unique in that no Crown forest tenures are renewable but all except timber licences, pulpwood agreements, and timber sale licences are replaceable every five to ten years by a new licence with a term equal to the licence it replaces.[7] If an offer of a replacement licence is rejected by the licensee, the existing licence remains in force until its term expires, at which point it is surrendered to the Crown.

The renewability of smaller tenures varies considerably by province and tenure type. Some, such as commercial cutting permits in Newfoundland and Labrador, are renewable. Others, such as timber supply licences in Saskatchewan, are renewable at the discretion of the minister, whereas some, for example, timber permits in Alberta and Manitoba and timber sale licences in British Columbia, are non-renewable.

Mill Appurtenancy

All provinces (with the exception of British Columbia, where all mill appurtenancy conditions were repealed in 2003) have some kind of appurtenancy requirements for, at least, their principal industrial forest tenures. Several provinces, including Alberta, Quebec, New Brunswick, and Newfoundland and Labrador, also have appurtenancy requirements for medium-sized tenures, and in some cases for smaller tenures.

Appurtenancy requirements vary considerably between the provinces in terms of the proportion of the AAC that is subject to appurtenancy statutes and/or regulations and the extent to which logs are directed through regulations to particular processing facilities. For example, some provinces, such as New Brunswick and Quebec, carefully regulate the flow of logs from all large and medium-sized tenures to specific process-ing plants. Others, such as Alberta, require licence holders to operate mills but do not direct log flows. Saskatchewan does not specifically require forest management agreement holders to operate a mill, but each owner of a processing facility must have a forest management agreement to provide access to a wood supply. In Manitoba, Ontario, Quebec, New Brunswick, Nova Scotia, and Newfoundland and Labrador, more than 80 percent of the provincial AAC is legally associated with designated wood-processing facilities.

Mutability and Compensation

In most provinces, governments reserve the right to modify tenures during their terms in the light of changing public objectives, changing land use imperatives, and new inventory information. Exceptions in-clude long-term timber licences in Newfoundland and Labrador and forest management agreements in Manitoba. For most long-term re-placeable or renewable tenures, adjustments to AACs and/or the areas under licence can be made at any time (e.g., British Columbia, Alberta, Ontario) or at regular intervals (e.g., Saskatchewan, Quebec, New Bruns-wick) or both.

Statutory provisions for governments to compensate tenure holders are usually restricted to land withdrawals but may include compensation for AAC decreases that result from reductions in areas under licence. Compensation usually includes, but is not restricted to, improvements made by licensees for which costs have not been recovered. In no province are procedures for calculating amounts of compensation by governments, in the event of land withdrawals or AAC reductions, set out in either statutes or regulations. Rather, the amounts seem to be discretionary or a matter of negotiation. In British Columbia, if agreement cannot be reached on amounts of compensation, disputes must be settled under the province's Commercial Arbitration Act.

In some cases (e.g., tree farm licences and woodlot licences in British Columbia and forest management agreements in Alberta), compensation is provided for AAC reductions beyond certain limits only – a statutory 5 percent of the AAC in British Columbia and a negotiated, usually about 3 percent of the net forested land base under licence in Alberta. In both provinces, compensation may be monetary or through exchanges of land.

Compensation may also occur between private companies. For example, in Alberta, oil and gas companies have rights to clear trees on forest tenures for exploration and energy extraction, but tenure holders are compensated. Unlike situations where the Crown pays compensation, holders of forest management agreements and energy companies have negotiated specific compensation schedules that dictate payment amounts.

Implications of the Characteristics of Crown Forest Tenure Systems for the Pursuit of Sustainable Forest Management

In the remainder of this chapter we explore current tenure arrangements in the context of sustainable forest management – a keystone management approach, as we have seen in the Introduction and Chapter 1, of all federal, provincial, and territorial jurisdictions in Canada. Specifically, to what extent do Crown forest tenures serve the goals of sustainable forest management, and what constraints do they present to the achievement of these goals? The discussion will be structured around the three generally recognized objectives of sustainable forest management: ecological, economic, and social. First, from an ecological perspective, policies should ensure that the structures of forest ecosystems are maintained in forms that will allow desired combinations of forest goods and environmental services to be produced now and in the future. Second, economic sustainability requires that forests be managed in ways that sustain, and ideally enhance, their contributions to incomes and employment – local, provincial, and national – over the long term. And, finally, the social component of sustainable forest management mainly has to do with the distribution of wealth generated by forests among individuals and groups within society and, particularly, among regions. Social sustainability is also increasingly concerned with gender equity and equal opportunities for visible minorities, with particular emphasis on First Nations. Each of the tenure characteristics described above will now be examined in the light of one or more of these three goals. Most tenure characteristics are not important in influencing all objectives of sustainable forest management. Below, we concentrate on those sustainable forest management objectives that are most directly related to a given tenure characteristic.

Initial Allocation of Tenure Rights

How tenures are allocated can have significant implications for the extent to which the goals of sustainable forest management are served. Two issues surrounding tenure allocations have the potential to influence the pursuit of sustainable forest management objectives. First is the degree to which the allocation of new tenures is competitive. More competitive processes that have open calls for tenders generally attract multiple applications, whereas less competitive processes are more closed, resulting in fewer applications. Second is the range of attributes that governments consider in adjudicating applications for Crown tenures. A narrow adjudication approach would focus on the amount of revenue that will be returned to the Crown, whereas broader criteria would consider non-pecuniary aspects, such as an undertaking that a specified plant will be built in a particular location, a certain number of jobs will be provided, or a prescribed approach to forest management be practised.

If allocated to the highest bidder through competitive processes that are solely price based, then an important aspect of economic sustainability will be served, since tenures will tend to be awarded to the most efficient operators that can generate the highest net revenues and contribute maximum amounts to public revenues. However, from a social perspective, governments may, and in fact frequently do, view Crown forest tenures as vehicles for delivering economic, social, and environmental values beyond pecuniary returns. Where such goals are pursued, tenures may be allocated through competitive processes, often involving multiple criteria, or, as is the case for a majority of Canadian Crown forest tenures (see Table 3.2), directly by provincial governments.

But direct awards may seriously compromise the pursuit of sustainable forest management, since they provide no assurance that rights are being awarded to that firm that can achieve public objectives most effectively. Without the transparency associated with competitive bidding, politicians may be subject to capture by specific interests at the cost of broader social interests. In fact, the criteria under which a tenure is awarded and the reasons a particular recipient is chosen may not be made generally available. Competition demands that the benchmarks against which bids will be assessed – whether this is price or a combination of price and other criteria – be explicit and widely advertised. The transparency of such a process engenders confidence in allocation procedures among both forest sector firms and the general public and helps motivate governments to pursue policies that reflect public aspirations.

Comprehensiveness

As we have seen, forest tenure arrangements in Canada, with some relatively minor exceptions, provide only the right to harvest timber, the rights to other resource values being either retained by the Crown or granted to other licensees in the form of overlapping tenures. Sometimes, a Crown forest tenure may not even grant complete rights to harvest all the timber. In some provinces, Alberta and Quebec for example, separate licences are provided to harvest hardwoods and softwoods from the same area or, in some cases, such as pulpwood agreements in British Columbia, sawlog harvesting rights are granted under one tenure arrangement and pulpwood harvesting rights under another for a common area. Such limited rights can have serious implications for sustainable forest management from ecological, economic, and social perspectives.

If rights to future forest crops are absent, reforestation and other silvicultural costs may be treated as current production costs that are weighed against the benefits of processing stocks of mature timber (Luckert and Haley 1993; Luckert 1998). Since there is no direct connection between the amount and types of silvicultural expenditures undertaken and the potential benefits they will produce in terms of the value of future stocks of timber on the land base to which they are applied, misallocation of funds, in terms of maximizing future net benefits, is inevitable, and the sustainability of timber supplies is likely to be compromised. This problem may be exacerbated by policies that invoke the "allowable cut effect." Under these arrangements, some provinces allow tenure holders to increase their allowable annual harvests of mature timber stocks if they can demonstrate that they have undertaken expenditures on silviculture that increase the growth rates of immature stands and thus the AAC of the management unit as a whole. Under such a policy, expenditures that generate increased future crops are rewarded immediately with benefits arising from the harvesting of mature timber in areas where species, timber quality, and harvesting costs could well be completely different from the sites where the silvicultural activities actually take place.

Increasing the comprehensiveness of property rights, it can be argued, will result in licensees practising forest management that is more eco-sensitive, since, given the rights to a broad spectrum of products, they will have an incentive to manage the forest landscape as an integral unit (H. Kimmins 2006). This contention is particularly significant if separate licences are granted to harvest different species, species types, or tree sizes. Under these circumstances, each licence holder will try to maximize net returns from one species, creating conflicts and possibly damaging the forest ecosystem and its future productivity (Hegan and Luckert 2000; Cumming and Armstrong 2001) . Coordinated management for a spectrum of forest products could also promote greater economic efficiency, resulting in the maximization of net benefits from the land base, and, since licensees will be sensitive to price signals across a spectrum of forest products markets, a socially desirable mix of goods and service may be produced (Haley and Luckert 1990).

On the other hand, in some cases, the benefits of specialization by separate rights holders may exceed the benefits to be gained by creating more comprehensive property rights arrangements. For example, there may be cogent arguments for separating surface rights to harvest forests from subsurface oil, gas, and mineral rights. Similarly, having two firms, one specializing in timber management and the other specializing in providing recreational services, might result in more effective multiple product management than providing either specialized firm with more comprehensive rights.

Rights to many non-timber forest products are not granted because it is recognized that such rights are difficult, if not impossible, to enforce from a practical, administrative perspective. The cost of controlling access to extensive areas of forestland for activities such as wilderness hiking, snowmobiling, backcountry skiing, or the collection of non-timber botanical products will often far exceed the potential revenues to be earned from such instruments as entry charges and permits. In such cases, it may be best to allow informal use of these resources without granting specific rights.

Trade-offs associated with benefits from specialization versus benefits from comprehensive and integrated management may be considered in market transactions. In the presence of well-functioning markets and clearly defined property rights to all resources, it is conceivable that rights negotiated among property right holders could result in the production of a socially optimum mix of outputs (Coase 1960). However, there are many instances where reliance on market solutions will fail to achieve social goals. For example, private coordination of forest management to produce multiple products will be efficient only if all the products concerned can be freely exchanged in the marketplace. But many forest products are not, and in some cases cannot (i.e., public goods such as biodiversity and the visual quality of landscapes), be bought and sold through conventional market channels.[8] Lacking the incentives of market signals, it is unlikely a private firm with rights to the products concerned will produce them in a socially optimal manner.

For instance, if a timber company is provided with rights to wildlife, it might decide to generate a stream of net revenue by selling hunting rights. However, in managing to maximize joint revenues from timber and hunting, it will likely concentrate on the production of large ungulates – deer, elk, and moose – to the potential detriment of other species of wildlife that have no market value or may even threaten the marketable species. Where non-marketed products comprise a significant and important component of the desirable product mix from a forest, a strong case can be made that the rights granted to tenure holders be limited and that coordinated resource management be a public responsibility.

In Canada, granting tenured rights to some forest products may also be constrained by cultural considerations. This is true of access to forest-land in pursuit of certain types of extensive outdoor recreation, hunting, and the collection of non-timber botanical products. Although in most cases, title to these products is retained by provincial governments, which have legal authority over their accessibility and the power to transfer them to third parties, many Canadians, particularly those living in rural areas, believe they have a customary right to such products, established through many years of unimpeded access at no charge. Although such usage does not, in itself, establish a customary right, it may be used to establish a legal claim to such a right. However, a successful challenge of this kind has never been mounted – at least by a non-First Nations plaintiff – in the Canadian courts. Nevertheless, this situation can create difficulties in extending tenured rights beyond timber. An example can be found in the case of British Columbia's community forest agreements. These agreements are the only type of tenure in Canada that grant the right to manage for and sell non-timber botanical products. However, any attempt by CFA holders to exercise this right has been met by local opposition and, to date, few attempts to control access and market these products have been successful.

Allotment Type

A serious problem facing forest policy makers intent on sustainably managing Crown forests through the tenure system arises when the tenures themselves are volume rather than area based. As we have seen, most provinces allocate the majority of their AAC to area-based tenures (see Table 3.3); however, an anomaly is British Columbia, where forest licences, which account for almost 60 percent of the province's allocated AAC, are volume based. Although volume-based timber tenures provide incentives to harvest trees by specifying an allowable annual or periodic harvest, they may fail to provide incentives for their holders to manage the resource sustainably over the long term. Since these holders occupy only the area in which they are operating fleetingly, possibly with little prospect that they will return to the same area in the future, licensees with volume-based tenures may have few, if any, incentives to consider what may be best for reforestation and stand management or even to harvest timber in ways that preserve the integrity of ecosystems or site productivity. Furthermore, licensees have little opportunity to accumulate useful site-specific knowledge concerning the areas in which they operate, and any knowledge they do gain may be lost when they move on to a new harvesting area. Under such arrangements, most planning and management activities become the responsibility of governments.

Having control over a geographically defined area may provide holders of area-based tenures a proprietary interest in the land under their control. They are readily judged by licensing authorities, their neighbours, and society at large on their management skills as reflected in the condition of their licensed area and, consequently, may have greater incentives than volume-based licence holders to take a long-term management perspective, conduct resource inventories, and accumulate the knowledge necessary for optimum management planning. However, if government and private objectives for forest management diverge, the government control inherent in volume-based allotments may be preferable. For example, the ability to direct harvesting in volume-based agreements in British Columbia allows the provincial government more options in attempting to control the current outbreak of mountain pine beetle that transcends tenure boundaries and the interests of any one tenure holder.

Size Restrictions

The achievements of ecological, economic, and social sustainability may all be affected by the size of the area within which tenured rights are granted. From an ecological perspective, a tenure should be large enough for it to be managed as an integral ecosystem or system of closely inter-related ecotypes. Forest managers often refer to such an area as a "land-scape unit." The optimum size of such a unit will depend on many factors, such as the ecological characteristics and variability of the forest, the drainage pattern in the area, and the topography. Restrictions on tenure size that lead to fragmentation of discrete landscape units may compromise sustainable ecosystem management.

From an economic perspective, efficiency requires that tenures be of a size that allows economies of scale to be realized. If property rights are freely divisible and transferable and free competitive markets for them exist, there will be a tendency for their size to adjust to a point where all private economies of scale are realized. This process is well recognized in the agricultural sector, although concerns, which may attract public intervention, are frequently raised about the negative social and distributive impacts of such adjustments – for example, the rise of corporate, industrialized agriculture at the expense of family farms and the demise of rural communities. However, where restrictions on divisibility and transferability of rights exist, which is common in the forestry sector, the size of tenures initially granted becomes an im-portant issue of economic policy.

Although a case can be made for tenures that are large enough to comprise a logical management unit from an ecological perspective and allow their holders to achieve economies of scale in management and production, from a social perspective they should not be so large that an undesirable degree of concentration of control over resources results. For example, the holder of a large Crown forest tenure that dominates a region may be able to exercise market monopsony powers by paying lower prices in its dealings with local suppliers and contractors than would occur in a competitive situation. Large holders may also enjoy a powerful negotiating position when dealing with local and provincial governments and may be the source of major regional economic and social disruption if they encounter business difficulties or face a major downturn in their markets.

In the Canadian provinces, as described above, there is no legislation that either limits the maximum size of major tenures or sets minimum sizes. Given that the optimum size of any tenure will depend on its purpose, location, and many other unique individual factors, it would be unwise to have such legal constraints. However, it is clear that holding size is an important consideration when designing a tenure system for the pursuit of sustainable forest management, and trade-offs may have to be made between ecological, social, and economic imperatives.

Exclusiveness

To the extent that exclusive rights are granted to an individual or well-defined group (i.e., common property), property rights holders will have sole access to the benefit streams from their properties and thereby have incentives to invest in improvements. Where benefit streams can be accessed by two or more individuals or groups, arranging who takes how much from streams of returns may be more complicated, and decisions regarding future investments will be more difficult to make, since each party will be aware that the other may capture more than its fair share of the returns. In the extreme, open-access provides no incentives to invest in projects designed to increase net benefit streams, as improvements will inevitably be captured by users – often referred to as "free riders" – other than the investor.

Canadian forest tenures, as we have seen, generally grant exclusive timber harvesting rights to each tenure holder, although there may be several exclusive rights granted to various products from a single piece of land. However, there are some forest resources to which exclusive rights are not, and cannot, be granted. For example, the benefit stream that occurs if a forest company maintains biological diversity may be enjoyed by all of society, not exclusively by the managing firm. Therefore, tenure holders have insufficient incentives to manage for such resources, and public regulation may be necessary in the social interest.

Transferability

Restricted transferability, which is a prominent feature of provincial Crown forest tenure arrangements in Canada, can have several negative impacts on the pursuit of sustainable forest management. Despite rotation lengths of seventy years or more between harvests of forest crops, incentives may exist for licensees to voluntarily invest in reforestation and the improvement of forest stands if expected net present values are positive and there are no barriers to the sale of their Crown tenures. However, if the opportunity to capitalize on their investments in timber production before crops reach maturity is constrained through restrictions on transferability, tenure holders must accept that their capital assets will remain frozen for many decades. Few investors are willing to relinquish liquidity to this degree, particularly when rates of return on their investments in timber production may be modest, as is frequently the case in Canada.

Perhaps a more important result of restrictions on the transferability of forest tenures, however, is that they impede their reallocation to those parties that are capable of using them most efficiently to maximize net values of the benefit streams they produce for their holders and their public owners. By restricting the extent to which the structure of the forest sector can adjust to changing markets, technology, and infrastructural developments, restricted transferability may place important constraints on the long-term economic sustainability of the forest sector.

Although a case can be made for freely transferable forest tenure rights, constraints on transferability may be justified on social grounds. Controlling the transfer of Crown assets permits governments to limit excessive, socially undesirable concentration in control over timber rights, forestall relocation of industrial activity that conflicts with regional employment and development objectives, and maintain a balance between domestic and foreign control over public resources.

Export Restrictions

Export restrictions on unmanufactured wood products from provincial Crown lands are generally justified on social grounds. It is argued that if, rather than being exported, logs are manufactured domestically into higher value products, more jobs are created in the provincial forest products manufacturing sector, helping to sustain forest-based communities. However, such controls may have serious negative consequences for sustainable forest management.

First, by restricting access to out-of-province log markets, log export controls limit log producers' market options and in some areas may significantly reduce local log prices to the benefit of domestic forest products manufacturers that, protected from out-of-province competition, may conduct inefficient operations.

Second, lower log prices reduce the value of standing timber, thereby decreasing Crown revenues from the sale of public timber, lowering the incomes of non-integrated licensees that are in the business of growing trees, and reducing the volume of timber that can be harvested at a profit. Thus, in addition to reducing processing efficiency, restrictions on log exports transfer wealth from timber owners – the public in the case of Crown forestland – to forest products manufacturers; discourage investments in timber growing by reducing prospective profits from such activities; reduce potential sustainable timber harvests by reducing the economically operable land base; and, while protecting some manufacturing jobs, decrease the number of workers involved in timber growing, harvesting, and exporting. In fact, it is possible that over the long term, removing log export restrictions could result in more jobs in the forest sector rather than fewer (Haley 2003).

Duration and Renewability

The term, or duration, of a forest tenure has important implications for its sustainable management. Any restrictions on term provide incentives to postpone costs and bring forward benefits. If rights to harvest timber extend over a long period, the holder is more likely to consider the relative merits of harvesting now rather than in the future. If rights extend to future timber crops, holders are more likely to ensure that current practices preserve soil productivity. Furthermore, incentives to invest in reforestation and other silvicultural treatments are more likely to take place if the duration of the right is sufficient to allow the returns from investments to be captured in the form of more valuable future timber harvests (Zhang and Pearse 1996).

Closely allied to the duration of a property right is the question of its renewability. If renewal, or replacement, of a right is guaranteed indefinitely under predetermined conditions, duration is less of an issue. However, if renewal is disallowed or the terms of renewal uncertain, duration is a matter of prime importance that may significantly affect the behaviour of tenure holders.

In spite of the disadvantages of shorter-term Crown forest tenures, they are often justified on the grounds that if rights are granted over longer terms, governments have less opportunity to modify tenure arrangements in the light of changing social values and public objectives.

Mill Appurtenancy

Mill appurtenancy requirements are generally created to ensure that logs are processed in the region in which they are harvested, thus maintaining local employment and stimulating regional development. However, such attenuations, although important components of policies designed to pursue social sustainability, may have negative implications for economic efficiency and industrial structure, and thus have negative consequences for the economic sustainability of the forest industrial sector.

Making the operation of a manufacturing plant a prerequisite for access to public timber creates a situation in which vertical integration is required. If logs from a tenure are inexorably linked to a particular manufacturing plant, as they are in some provinces, firms can neither allocate them to their most profitable use in other mills within the company nor trade them with, or sell them to, other firms that can use them more efficiently. Consequently, with vertical integration, log markets are likely to be thin or absent. In this way, appurtenancy requirements hinder the creation of open, regional log markets where logs can be bought and sold competitively and thus allocated to their highest and best use. The lack of developed log markets also has implications for stumpage systems, which frequently rely on prices of processed forest products rather than logs. These issues are discussed in Chapter 5.

Appurtenancy requirements may also influence forest management. Forced vertical integration, which ties specific forests to specific mills, is the root cause of problems associated with coordinating mixed-wood management activities, discussed above. As technology has created new uses for some species of trees previously considered to be weeds, the original holder of the forest tenure may not have the plant necessary to process the newly commercialized species. Thus, the rights to these species are frequently awarded to another tenure holder, thereby creating overlapping tenures.

The forced vertical integration of appurtenancy requirements also creates barriers to entry. Once available timber has been allocated, as is generally the case in all provinces, entry to the industry is severely constrained by appurtenancy requirements, since, in the absence of log markets, the only avenue for a would-be entrant to pursue is to acquire an existing company's harvesting rights. Firms within the industry can grow only by acquiring competitors' harvesting rights through purchase, merger, or takeover. Thus, control over regional wood supplies may become highly concentrated.

The forced vertical integration that results from appurtenancy requirements influences the business structure of firms as well. Since log trading is constrained, log production is not a profit centre in companies that have to conform to appurtenancy requirements. Logs are transferred to manufacturing plants at costs that may be below fair market price. Consequently, timber production subsidizes manufacturing – the corporate profit centre – with cheaper-than-market inputs, and a major incentive for manufacturing plants to be efficient is removed (Haley and Luckert 1991).

Finally, appurtenancy requirements can reduce the flexibility of firms to respond to changing market conditions. For example, appurtenancy requirements could cause wood supply to be forfeited if a plant is closed or relocated. The threat of losing their wood supplies makes it difficult for firms to rationalize their operations by taking advantage of changing transportation infrastructure and technology or concentrating production to achieve the economies of scale necessary to compete in global markets.

Mutability and Compensation

Mutability, by creating uncertainty, may reduce the amount of capital invested in forest management by tenure holders and thereby diminish the prospect that forest resources will be managed sustainably. However, if tenure holders know in advance the degree to which tenures can be legally modified, this element of uncertainty can be more confidently incorporated into management decisions.

The uncertainty associated with mutable tenure arrangements can be mitigated if contractual provisions exist for the payment of compensation to the tenure holder should the government decide to modify the conditions under which the tenure is held or even cancel the arrangement in whole or in part. If the conditions surrounding compensation are ambiguous and/or appear arbitrary, as is frequently the case in Canada's Crown forest tenure system, the capacity to mitigate uncertainty will be much less than if the terms under which compensation will be paid are clearly spelled out. Clarity with respect to how amounts of compensation will be determined and the conditions under which they will be awarded allow tenure holders to plan more rationally for the risks that mutable tenures present.

Of course, not all changes that reduce the profitability of tenure holders' rights should necessarily be compensated. Just as tenure holders' investments are made while considering political risks and uncertainties associated with market prices and costs, it is also important that such investment decisions consider risks associated with legal changes to their tenures.[9] For example, if legislated compensation completely insulates a tenure holder from the prospect of increasingly stringent forest practices regulations designed to protect forest ecosystems, they will have no incentive to plan for flexibility in meeting changing social expectations.[10] The key is for government policy changes to be well conceived and justified so that arbitrary decisions do not create an insecure investment environment.

Conclusion

Provincial Crown forest tenures in Canada differ in many ways that reflect their history, purpose, and the socio-political environments in which they have evolved, but they also have much in common. First, with minor exceptions, they provide exclusive rights only to harvest timber – sometimes of a specific species or type – and do not extend rights to other forest products – in most cases, not even to the productivity of the land itself. Second, with the exception of British Columbia, timber harvesting rights are mainly allocated by means of area-based tenures. Third, again with the exception of British Columbia, forest tenures cannot be transferred without permission, usually from the provincial minister responsible for forests. Fourth, most provinces control the export of unmanufactured forest products from Crown forestlands, but no province has an outright ban on such exports. Fifth, the terms of most major forest tenures are very similar – twenty to twenty-five years – and fall short of tree crop rotation lengths. Arrangements for renewal (replacement in British Columbia) are generally "evergreen" on a five- to ten-year basis. Finally, except in British Columbia, most major Crown forest tenures in Canada are issued with a proviso that the licensee must own and/or operate a manufacturing facility.

Some aspects of Canada's provincial Crown forest tenure system may place constraints on the attainment of sustainable forest management. However, in many cases, tenure arrangements negatively impact the realization of some sustainable forest management goals while positively serving others. For example, mill appurtenancy and controls on transferability may pose serious barriers to economic sustainability while serving social sustainability by helping maintain local jobs, sustaining the economic health of forest-dependent communities, and allowing governments to influence the structure of the forest industry and forestall undesirable concentration of industrial influence on regional economies. Moreover, tenure terms that fall short of rotation lengths may be undesirable from the perspective of ecological sustainability but provide governments with the opportunity to adjust tenure arrangements more frequently to better reflect changing social values and objectives. Clearly, when devising a tenure system that best promotes forest sector sustainability, there are many trade-offs to be made between the often conflicting impacts of key tenure characteristics on the various goals of sustainable forest management. These matters are explored in more depth in Chapter 6.

Two important dimensions of Canada's Crown forest tenure systems have yet to be examined: forest practices and Crown stumpage arrangements. These issues are addressed in Chapter 4 and Chapter 5, respectively.

Regulating for Sustainable Forest Management: Interprovincial Comparison of Forest Planning and Practice Requirements

4

Since the 1970s, Canadians have been increasingly concerned about protecting environmental values in forests, including scenic quality, water quality and quantity, fish and wildlife, and biodiversity more generally. As a result, forest tenures have been increasingly attenuated by operational controls designed to protect a wide range of environmental values. These restrictions were often originally put in tenure documents, but over the past several decades, provincial governments have developed separate regulatory legislation to govern forest planning and practices (Ross 1995; Howlett 2001b).

Forest planning and practices regulations have formed the centrepiece of forest sector policy strategies across Canada to ensure that forest management addressed the environmental component of sustainable forest management. However, the complexity and rigour of these standards have created a significant cost burden and constraint on management flexibility, creating challenges for the economic component of sustainable forest management. As a result, these requirements have received increased criticism from industry and political parties oriented toward free enterprise. According to economic theory, private firms would not be expected to produce the optimal level of values in these areas because they are external to the costs and benefits that they receive, and may exhibit public good properties. These properties create a clear justification for some type of government regulatory role. But increasingly, questions have emerged about whether the style of "command and control" regulation is best suited to pursuing sustainable forest management (Pearse 1998). British Columbia has probably gone the furthest in pursuing reform to forest planning and practices regulations, and its experience reveals the challenges in finessing the trade-offs between environmental and economic sustainability.

This chapter begins with an overview of the issues involved in the optimal design of forest planning and practices policies. We then turn to analyzing the specifics of approaches used in the five largest timber-producing provinces.

Challenges in Designing Regulations to Protect Forest-Related Environmental Values

There are several different approaches to forest practices regulation, reflected in different jurisdictions around the world. Four general approaches are outlined in Table 4.1. First, guidelines can be used to identify recommended practices. In this case, the standards are not legally binding, and operators cannot be penalized for not adopting them. This approach is used in some southeastern states in the United States (Cashore and Auld 2003). Second, technology- or practices-based regulations specify particular forest practices that must be used in certain circumstances. This approach is also frequently referred to as "prescriptive." An example would be a thirty-metre buffer strip on a fish-bearing stream. Third, performance- or results-based regulations specify an outcome to be achieved rather than a specific practice. In the case of stream protection, an example would be maintaining water quality within the natural range of variation. The final approach to regulating forest policy is compulsory management planning, which requires operators to prepare a management plan but does not specify any particular practices or results that must be achieved. As such, this approach concentrates on input requirements as opposed to outputs. Unlike guidelines, practice- and results-based regulations and compulsory management planning are legally enforceable.

These four approaches are not mutually exclusive. Indeed, a number of combinations are possible. Policy makers might choose to have legally

Table 4.1 Alternative approaches to regulating forest practices

Approach	Example from protecting riparian values
Guidelines (best practices)	suggest, but do not require, practices like stream buffers
Technology- or practice-based regulations	30-metre no-harvest zone
Performance- or results-based regulations	maintain water quality within the range of natural variation
Compulsory management planning	requirement to develop a plan to protect riparian values

binding practices or results for some values or situations but rely on guidelines for others. Management plans could be required to meet certain performance requirements. There are various ways that these approaches can be combined in a single regulatory regime.

The challenge for policy makers is to design the most appropriate approach, or mix of approaches, for the conditions they face. The optimal choice will depend on numerous factors (Coglianese and Lazer 2003; P. May 2007). When objectives are easily measured, performance-based regulations are desirable. Since operators usually have superior information about how best to achieve a particular result, performance standards can be more cost-effective. When objectives are not easily measured, however, policy makers are more likely to rely on practices regulations that they believe will adequately protect the value of concern. However, when the problems confronting managers are highly diverse, uniform practices or performances regulations are likely to be ineffective. In this case, compulsory management planning might be the best alternative. It allows operators to tailor forest practices to distinctive local circumstances.

Perhaps the most vexing challenge in designing any forest practices regime is how to provide for variability over space and time. One of the great dilemmas in forestry is tailoring policy design to the exceptionally varied and complex nature of the problem. J.P. Kimmins writes that it was "the Great Mistake" of earlier stages of forest management "to ignore the spatial variability in the ecological characteristics of the forest" J.P. Kimmins 2000, 16). Uniform rules that apply over a jurisdiction may be clear and easy to understand, but they will not be effective or efficient if the circumstances in which they apply vary significantly. Policies designed to account for the variation in ecological, economic, and social conditions across a jurisdiction may be more effective in some ways, but their complexity may make it harder to assure the public that the values they want in the forest are being protected. Adapting to new information or conditions that emerge over time is also a difficult task.

Finding the optimal precision of rules – a phrase developed by law scholar Colin Diver (1989) – is a difficult design challenge. Diver describes three criteria for optimal precision:

- Simplicity is the number of steps involved in the decision rule.
- Transparency is the clarity of the rule: in the same situation, different people will interpret the rule in the same manner.
- Congruence is how well the design of the rule matches the problem it is intended to address. A congruent rule will not "over include" by being applied in situations where it is inappropriate, nor will it "under include" by failing to be applied in situations where it should.

The major dilemma is how to design rules that match exceptionally diverse problems while simultaneously being simple and clear. Three alternative approaches to addressing the issue of rule congruence are described below:

- *Prescriptive congruence* increases the complexity of the rules so that they vary to take into account all possible permutations. Although improving congruence, this approach departs from simplicity and may be administratively impractical to develop effectively, and very cumbersome to implement.
- *Professional delegation* relies on professional discretion by delegating significant authority to foresters and other professionals to adopt the practice they deem to be appropriate in the circumstance. This approach has the advantage of simplicity, but it sacrifices transparency.

- *Geographical delegation* decentralizes authority to lower levels of an organization to decide how to tailor solutions to local situations. This approach is more complex than professional delegation, but depending on how it is designed, could be more transparent. (Hoberg 2002)

In trying to strike this balance, policy makers need to take into account environmental effectiveness and economic efficiency, but also social and political criteria such as legitimacy and accountability (Gunningham and Grabosky 1998; P. May 2007). Sometimes, there are clearly trade-offs between economic and environmental objectives. We will see how the approach taken by Canadian governments to address the environmental component of sustainable forest management has placed considerable stress on the economic component. There are also potentially significant trade-offs between economic objectives and socio-political objectives. For example, policy designs that rely on professional discretion may be simple and potentially cost-reducing, but they may lack transparency, thereby creating significant problems with accountability. The dilemma is how to provide for the flexibility needed to address the tremendous variation of circumstances across the forested landscape while simultaneously ensuring environmental values are, demonstrably, protected. As we will see in the remainder of this chapter, efforts to move away from prescriptive planning and practice requirements have been constrained by the political need to demonstrate commitment to the protection of environmental values.

Our framework for analyzing rules for forest planning and practices has four categories: strategic planning, operational planning, forest practice requirements, and compliance and enforcement. Strategic plans address the higher-level plans that establish the objectives for resource management, frequently through the designation of differentiated zones across the relevant land base. Operational plans involve more site-specific development and approval for forestry activities, such as the specific silvicultural prescriptions and the location of roads and cutblocks (Davis et al. 2000; Nelson 2004). For both of these levels, our analysis will examine the content of plans, the process for the development and approval of the plans, and whether the plans are binding on lower-level plans or forest operations.

We also analyze specific standards for forest practices in the following areas:

- reforestation and other silvicultural operations,
- riparian buffers – the extent to which water courses must be protected,
- ecological requirements – spatial regulations, including opening sizes and green up/adjacency constraints; representation/stand structure requirements.

The impact of any set of rules is heavily influenced by the mechanisms established for compliance and enforcement. We review who is responsible for inspections and any requirements for inspections and audits, the system of penalties for non-compliance, and mechanisms for review and appeal of enforcement actions.

We have surveyed the forest practice and planning rules of the five most important provinces with respect to forestry: British Columbia, Alberta, Ontario, Quebec, and New Brunswick.

Results of the Comparison

Strategic Planning

All five provinces have a framework for strategic planning. The most common approach is to require long-term strategic management plans as part of a condition for holding a forest tenure. Content requirements for plans are laid out in planning manuals that have a quasi-legal status. As a result, strategic planning in Canadian forestry is pursued through compulsory management planning delegated to geographic locales specified by the boundaries of the tenures. The exception to this general pattern is British Columbia, which because of the dominance of volume-based tenures there adopts a different approach.

In Alberta, as a condition of their licence, holders of forest management agreements are required to develop detailed forest management plans (FMPs) for their forest management unit. Content and procedural requirements for the plans are specified in the *Alberta Forest Management Planning Standard* (Alberta Sustainable Resource Development 2006a). The plans have a ten-year term but adopt a two-hundred-year planning horizon, and include a stand-level spatial harvest sequence plan and map depicting operable forest stands scheduled for timber harvesting in the next twenty years. The forest management plans must address resource management philosophy, resource management goals (biological, economic, and social), forest management objectives and strategies, implementation strategies, and performance monitoring. In 2006, the province required the plan to explicitly address all components of the Canadian Standards Association's forest certification system.[1] Allowable annual cuts are established through FMPs.

In Ontario, FMPs need to be prepared for each of the province's forty-eight forest management units by companies holding sustainable forest licences. Planning requirements are described in the *Forest Management Planning Manual* (Ontario Ministry of Natural Resources 2009). The FMP must, in the words of the Crown Forest Sustainability Act, "describe the forest management objectives and strategies applicable to the management unit," and must have "regard for plant life, animal life, water, soil, air and social and economic values, including recreational values and heritage values."[2] The plan must be certified by a professional forester and receive approval of the minister of natural resources. It has a ten-year term and ten-year planning horizon. Unlike in other provinces, annual harvest levels in Ontario are determined on the basis of area, not volume. Available harvest areas are determined for each forest management unit as part of the FMP process.

In Quebec, tenure holders must submit a general forest management plan (GFMP) every five years (Quebec Ministry of Natural Resources, Wildlife and Parks 2007). If there is more than one tenure holder in a common area, a joint plan must be filed. Forest engineers (Quebec's term for professional foresters) must certify these plans. These are strategic plans that include a description of the management unit, AACs, and management objectives and strategies. Starting with the 2007-12 term, licensees must demonstrate how their GFMPs meet eleven protection and development objectives (PDOs) related to the protection of soil and water, biodiversity, and socioeconomic values. The PDOs were established in 2005 by the minister of natural resources, wildlife and parks, after a period of public consultation (Quebec Ministry of Natural

Resources, Wildlife and Parks 2005). The plans must be approved by the minister.

In New Brunswick, each of the province's ten timber licence holders must create an FMP specifying how it plans to meet government goals, timber and non-timber objectives, and standards (New Brunswick Department of Natural Resources 2006). FMPs cover a twenty-five-year period, are reviewed and updated every five years, and have a planning horizon of eighty years (Martin 2003). A management plan must include a description of objectives and how the licensee will manage the land with respect to silviculture, harvesting, road construction, fish and wildlife habitat, and other concerns. FMPs must be prepared by, or under the supervision of, a professional forester. The public must be consulted during the plan's elaboration. The minister of natural resources is responsible for approving the plan. The annual allowable harvest is assigned to licensees and sub-licensee mills through the forest management agreements for each year of the five-year agreement period.

British Columbia is distinctive, in large part because of the dominance of volume-based tenures in that province, which make strategic planning for areas more challenging. British Columbia does have a sophisticated land use planning process, based on land and resource management plans (LRMPs) and more landscape-level sustainable resource management plans (SRMPs) (Frame, Gunton, and Day 2004; Mascarenhas and Scarce 2004). Although LRMPs or other strategic land use plans are completed for 85 percent of the province, many of them do not contain sufficiently specific management targets to guide forest operations. The more specific SRMP process is intended to refine LRMP objectives at the small to medium landscape and watershed level (such as identifying old-growth management areas). Of the 195 SRMPs, 102 have been completed and 93 are underway (BC Ministry of Agriculture and Lands 2006). However, few of these contain legally established objectives (BC Forest Practices Board 2004).

The large-area-based tenures in the province, tree farm licences, do require detailed strategic management plans that require inventories, timber supply analyses, and specification of management objectives and strategies. These management plans are developed by licensees and approved by the chief forester. However, tree farm licences account for only 21 percent of the annual harvest in the province, and there are no similar strategic management plans required for the remaining 79 percent of the annual harvest.

The lack of strategic FMPs in volume-based tenures has been recognized as a problem for some time, and efforts were made to address the problems when the Campbell government came into power in 2001.[3] Rather than address the issue of tenure allotment type directly, however, the government proposed the "defined forest area management" approach, which encouraged volume-based licensees to cooperate on area-based sustainable FMPs in exchange for access to funds from the Forest Investment Account. However, the initiative was dropped after industry bridled against an additional planning obligation.

Operational Planning
Like the higher-level strategic plans, operational planning occurs through geographically delegated compulsory management planning. All five of the provinces examined have detailed operational planning requirements. In Alberta, operational planning occurs through annual operating plans and general development plans that must be reviewed and approved by the government. The details and requirements for these plans are specified in the "Ground Rules," formally known as the Alberta Timber Harvest Planning and Operating Ground Rules Framework for Renewal (Alberta Sustainable Resource Development 2006b). The general development plan provides a detailed description of a harvesting and reclamation strategy of the timber operator for a five-year period, summarizing deviations from the spatial harvest sequence described in the FMP. The timber operator must submit the general development plan every year by 1 May to the government, which must respond within sixty days. Forest management agreement holders need to file an annual operating plan that indicates how timber harvesting will be carried out, including how, where, and when the operator will develop roads, carry out harvesting and integrate operations with other forest users, mitigate logging impacts, and carry out reforestation. A compartment assessment may be required for a specific area within the forest management unit if the government determines that information or major issues are identified that have not been addressed in the FMP or spatial harvest sequence.

In Ontario, licensees must prepare forest operations prescriptions and also annual work plans (Ontario Ministry of Natural Resources 1995). The forest operations prescription is "a site-specific set of harvest, renewal and maintenance activities that will be used to ensure that the current forest is managed to achieve the expected forest structure and condition" (Ontario Ministry of Natural Resources 2009, D-5). The prescription must be certified by a professional forester. The annual work schedule, which describes areas to be harvested, road developments, and so on, also needs to be certified by a professional forester and approved by the Ministry of Natural Resources district manager (Ontario Ministry of Natural Resources 2009, part D).

In Quebec, agreement holders must submit an annual forest management plan (AFMP) for every area from which they are authorized to cut wood. As with the strategic general forest management plans (GFMPs), joint filing is required where there is more than one tenure holder in a given area, a forest engineer must approve the plan, and the plan must be submitted to the minister for approval. Beginning with the 2007-12 term, the AFMPs must demonstrate progress toward meeting targets set for the protection and development objectives (PDOs) contained in the GFMPs. The PDO implementation document notes that "once the GFMP has been approved, all subsequent annual plans must provide for effective application of the actions proposed. If the actions are not carried out, the MRNFP may use its power to order their implementation" (MRNFP 2004, 68). As the AFMP must be consistent with the GFMP, any tenure holders that wish to engage in an activity not detailed in the GFMP need to prepare a new GFMP together with the same parties involved in preparing the original plan. A management permit is issued only upon minister approval of the AFMP.

In New Brunswick, annual operating plans must be submitted to the minister of natural resources in order to demonstrate how licensees intend to carry out their five-year management plans in practice. The operating plans are revised every year and must describe how licensees will meet government objectives and operating standards. Operational plans must be prepared by, or under the supervision of, a professional forester and approved by the minister of natural resources.

British Columbia's framework changed with the introduction of the Forest and Range Practices Act in 2004. Forest development plans, used under the 1994 Forest Practices Code, have been replaced by forest stewardship plans. These plans have a five-year term and must contain a map that locates cutblocks and roads. The results-based component is that licensees can propose results and strategies to meet objectives specified in regulation. The plan needs to be prepared by a professional forester and submitted for approval to the district manager of the Ministry of Forests and Range. Site-level silviculture plans need to be completed but do not need to be submitted for government approval. Unlike the other four provinces surveyed here, British Columbia does not require operating plans to be reviewed and approved on an annual basis.

Specific Forest Practices
The five provinces surveyed regulate forest practices through a complex mix of compulsory planning requirements and specific regulations. For the most part, the tendency is for regulatory standards to be practice based rather than results based, although British Columbia has made a concerted effort to move in a results-based direction. Four out the five provinces have forest planning manuals that provide direction to the planning process, and some (such as Ontario's) are supplemented by specific regulations. The exception, again, is British Columbia, which has no such manuals but relies on a regulation instead. The implications of these differences are not as significant as they seem, however, since the manuals take on a de facto regulatory role (Ontario's, in fact, are referred to as "regulated manuals") when the standards specified in the manuals are incorporated in approved management plans. The result is a similar structure of forest practice requirements, and how they relate to planning requirements. All are based on a structure of various practices regulations, with the option to vary from those regulations if provided for in an approved plan.

For example, British Columbia recently introduced a new Forest and Range Practices Act (S.B.C. 2002, c. 69) designed to simplify planning requirements and provide a more "results-based" operating framework.

The core regulation, the Forest Planning and Practices Regulation, contains numerous standards for a series of objectives. There are clear-cut sizes designed to protect biodiversity, and streamside buffers to protect riparian values. But licensees can vary from these so-called default standards if they can convince statutory decision makers that their alternative approaches can attain the objectives provided for in regulations (Association of BC Forest Professionals 2005).

The structure of the BC approach can be seen by examining the limits on clear-cut size. The Forest Planning and Practices Regulation specifies the following objective for landscape-level biodiversity: "Without unduly reducing the supply of timber from British Columbia's forests, and to the extent practicable, to design areas on which the timber harvesting is to be carried out that resemble, both spatially and temporally, the patterns of natural disturbance that occur within the landscape" (B.C. Reg. 14/2004).

The regulation provides specific limits to opening size: forty hectares in coastal and southern Interior regions, and sixty hectares in northern interior regions. But licensees can choose a different approach if they can convince the district manager in their forest stewardship plan that their approach meets the general objective stated above.

Alberta's Ground Rules place emphasis on abiding by a ten-year spatial harvest sequence laid out in the FMP, without mandatory block size constraints (Alberta Sustainable Resource Development 2006b). However, in the absence of a spatial harvest sequence, a preliminary harvest plan is required, and detailed spatial constraints apply. For example, the size of clear-cuts in pine and deciduous areas is stated as follows: "Harvest areas in deciduous stands or in stands where pine comprises 40% or more of the merchantable timber volume (evenly distributed throughout the harvest area) may be up to 100 hectares in area unless otherwise approved by Alberta, but shall average no more than 60 hectares" (Alberta Sustainable Resource Development 2006b, 29).

Additional requirements apply to both spatial harvest sequence and preliminary harvest plans, such as minimizing line of sight from roads to under four hundred metres, and ensuring that distance to wildlife hiding cover is under two hundred metres.

In Ontario, many standards are outlined in the *Forest Management Planning Manual* (Ontario Ministry of Natural Resources 2009), but variances can be obtained through their Forest Management Plans.

In the case of reforestation, all five provinces have mandatory reforestation requirements. Stocking standards vary by species and ecological classification, and the provinces have guidebooks of varying levels of detail that help inform the specific stocking standards provided in operational plans for particular sites. In the case of protection of riparian values, all five jurisdictions have practice-based approaches that rely on special management zones varying by watercourse classification.

In the case of clear-cut sizes, the five provinces all regulate opening size and allow it to vary according to regional characteristics. As suggested above, there are significant differences in maximum clear-cut sizes (Cashore and McDermott 2004), from 24 hectares in the spruce zone in Alberta to 260 hectares in Ontario. All five jurisdictions also have regulations that require leaving some elements of forest structure behind during harvesting, but they are designed quite differently (Hoberg and Karmona 2005).

One interesting difference between British Columbia and other provinces is how economic impacts are considered in the design and implementation of forest practices regulations. British Columbia has a very specific policy that limits the impact of its forest practices regulations such that these regulations may not reduce an allowable annual cut by more than 6 percent – and this 6 percent is subdivided into specific components of wildlife, riparian, stand-level biodiversity, and so on (Hoberg 2001a; BC Forest Practices Board 2004). Indeed, each of the government's objectives for forest practices in the Forest Planning and Practices Regulation is explicitly qualified in the regulation by the rather prosaic phrase "without unduly reducing the supply of timber from British Columbia's forests."[4] None of the other provinces is so explicit in the constraints placed on forest practices.

Compliance and Enforcement

All five provinces have systems of inspection and authorize the use of administrative and criminal penalties to penalize non-compliance. No comparative studies exist of degrees of compliance or actions of enforcement officers. The principal differences among the provinces are the differences in reliance on self-inspection. Both Alberta and Ontario rely to a large extent on industry self-inspection (Winfield and Benevides 2003). Quebec has a system of government inspection. British Columbia has the most intensive government-centred framework. It supplements its tens of thousands of annual inspections by audits by the quasi-independent Forest Practices Board. In Ontario, there are independent, non-governmental audits every five years, which must be submitted to the legislature. The Ministry of Natural Resources and licence holders jointly develop an action plan to respond. New Brunswick also requires audits every five years.

Patterns from the Comparison
This review of forest practice requirements reveals an extremely complex mix of regulatory requirements. But there is a great deal of commonality in the structure of forest planning and practice requirements across the five provinces surveyed. Each province has formal, compulsory requirements for planning and a framework of regulations to govern forest practices. For the most part, these standards are practice standards. British Columbia has made a deliberate shift toward a more results-based framework, but its progress toward that goal will be difficult to judge until more forest stewardship plans have been developed and approved. All five provinces practice a similar approach to congruence, or how rules are designed to vary to match different problems. They combine prescriptive congruence, which builds variation into regulations by creating different rules for different conditions, with geographical delegation, where variations are built into geographically specific management plans or exemptions by statutory decision makers.

There are differences in the settings of some regulatory requirements, such as clear-cut sizes. The provinces also allocate responsibilities for inspections differently, with Alberta and Ontario providing a greater role for industry self-inspection. But perhaps the most significant differences between the provinces are the small fraction of the BC forestland base that is covered by requirements for strategic forest management planning. This difference results from the predominance of volume-based tenures in British Columbia. Only the large-area-based tree farm licences, which make up only one-fifth of the allowable cut from Crown lands, are covered by requirements for strategic management plans. This does not mean that strategic planning is absent on the remaining land base. British Columbia has a government land use planning process that provides some strategic guidance. Moreover, the chief forester uses an elaborate process to set allowable annual cuts throughout the province. Nonetheless, there are concerns about the quality of strategic planning on much of the land base, particularly the paucity of legally established objectives. Further research is required to examine the consequences of the absence of more compulsory strategic planning in British Columbia.

Implications of Planning and Practice Requirements for the Pursuit of Sustainable Forest Management

The intensification of forest planning and practices requirements reveals the tensions within the sustainable forest management paradigm. Operational rules for forest practices have emerged in response to increasing attention to the environmental component of sustainable forest management. But these efforts have created additional strains on the ability of the sector to meet economic objectives.

The intensification of requirements has certainly affected Canada's environmental performance. In a global context, Canadian provinces have been shown to adopt a relatively stringent approach to regulating forest practices. Cashore and McDermott (2004) compare forest practices in thirty-six jurisdictions around the world. They developed a stringency rating that was based on the extent to which rules were mandatory and contained substantive (as opposed to procedural) restrictions on activities. Countries were arrayed in a matrix with a scale from one to nine, with nine being the most stringent. British Columbia and Alberta both received the highest stringency rating of nine, and Ontario and Quebec were just one notch down with a stringency rating of eight (Cashore and McDermott 2004, ch. 7, 43). (The study did not examine New Brunswick.) These scores do not speak to the efficacy or efficiency of forest practices regimes, but they do confirm that, in a comparative context, Canadian provinces tend to take a rigorous approach to the protection of environmental values in the forest.

Although many environmentalists remain critical about the environmental performance of the Canadian forest sector, industry officials, professional foresters, and policy analysts have expressed concern that the framework of environmental rules is unduly complex and burdensome, resulting in economically inefficient policies that undercut industry competitiveness. Thus far, there has been remarkably little innovation in the design of environmental regulatory requirements in Canadian forestry. The BC government has tried to address this concern most directly, introducing a results-based code, formally known as the Forest and Range Practices Act, designed to maintain environmental standards while reducing the complexity and costs and increasing operational flexibility of the policy framework.[5] One significant change in their framework was the elimination of approval for site-level plans, which has reduced the administrative complexity of the system significantly and increased operational flexibility.

However, there was not as much change in the regulatory structure as many in industry had hoped (and as many in the environmental community had feared). The government found it extremely difficult to develop measurable results-based standards, and as a result, many of the practices and standards from regulations and guidelines in the old Forest Practices Code have been incorporated into the new regulations. What is different, however, is the degree of flexibility licensees have to propose alternatives to these default rules. The new regulatory scheme creates more emphasis on licensees' proposed results and strategies to address regulatory objectives in their forest stewardship plans – a specific attempt to move from prescriptive standards to a form of geographically specific compulsory management plan. How this all plays out in terms of the balance between environmental protection and economic production will be revealed only as the new regulatory framework is implemented over time.

The early results from the implementation of the new framework suggest that the expectations that licensees and their forest professionals would use the new-found flexibility to pursue innovative new approaches were unrealistic. A report of the Forest Practices Board reviewing fifteen early forest stewardship plans concluded that licensees were not committing themselves to measurable results but instead relying either on the default standards from the old code or on strategies without measurable outcomes (BC Forest Practices Board 2006).

One lesson of the BC efforts to reform forest practices is how constraining the political imperative to ensure the effectiveness of environmental protections in the forest is. British Columbia has managed to infuse greater flexibility into its system, but it is still highly prescriptive. Performance-based and compulsory planning approaches promise greater flexibility and the potential for greater cost-effectiveness, but governments have been reluctant to rely on them because of their perceptions of the political risks involved in reducing the apparent certainty of achieving specific outcomes. This dynamic also appears to have operated in Alberta, where the province has moved over the past decade from an objective-based approach to a far more prescriptive one (Golec and Luckert 2008). This political imperative for environmental protection is also revealed in the recent efforts by other provincial governments – Ontario and Quebec – to update their forest planning frameworks. Overall, the effect of the new standards in these provinces has been to increase the force of operational rules constraining forest managers.

Finally, it is striking how little comparative information is available about the force and effect of the different regulatory systems across Canada. This chapter – and other works, such as Cashore and McDermott (2004) – has examined the framework for operational rules outlined in statutes, regulations, and planning manuals, but there has been no analysis to date that has assessed the comparative impact of these regulations on the social, economic, and environmental values promoted by the concept of sustainable forest management.

Interprovincial Comparison   5
of Crown Stumpage Fee Systems

One of the most contentious issues in forest management is how to estimate the value of standing trees or stumpage fees.[1] In cases where markets exist for standing timber, stumpage prices may be determined through market transactions. However, in Canada, where forestlands are predominantly publicly owned and exclusive timber harvesting rights are granted to private companies over lengthy periods through Crown tenure systems, the question of how to assign stumpage values takes on special significance. Accordingly, stumpage fee systems attract the intense and continuous attention of forest policy analysts and provincial policy makers.

Stumpage fees are the most important component of Crown tenure holders' fiscal obligations and can comprise a major attenuation of their rights. Stumpage payments may account for a significant proportion of a manufacturing plant's delivered wood costs and have impacts on tenure holders' behaviour that may be of considerable social significance (Luckert and Bernard 1993; Haley 2001).[2] First, stumpage fees act as important signals to forestry firms that may influence their timber growing, harvesting, and processing strategies. If stumpage fees do not reflect the true social value of standing timber, decisions taken by firms will be distorted from a societal perspective. For example, stumpage fees set below the true social value of timber may discourage firms from investing in wood-saving processing technology and reduce incentives for reforestation and the management of timber crops. Second, stumpage fees may have important impacts on the mix of products produced from a forest area, since, when trade-offs have to be made, they are an important indication of the benefits of producing timber compared to the production of non-timber goods and/or services. Finally, stumpage prices may have serious consequences for international trade relations. Canada's long-standing softwood lumber trade dispute with the United States largely hinges on the Americans' contention that the lumber industry in Canada is being subsidized by means of administered stumpage rates that are lower than they would be if determined in competitive markets.[3]

Stumpage fees in Canada are regarded not simply as the price the private sector pays in order to acquire public timber but as important instruments of forest policy that can be used to pursue diverse social objectives, such as regional development, economic stability, or the mix of timber products being manufactured. In fact, the great variety of provincial stumpage fee systems seen across Canada is a reflection of the varied objectives that provincial governments are trying to achieve (Haley and Luckert 1990).

In this chapter, following a theoretical discussion of the principles of determining standing timber values, we address the inherent problems of translating these principles into practical and equitable stumpage fee systems that meet the public purpose. A framework for comparing stumpage fee systems across Canada is then developed, and this scheme is used to describe methods of determining, collecting, and dispersing stumpage fees in the Canadian provinces, excluding Prince Edward Island. Finally, provincial stumpage fee systems are compared, and the issues they raise in relation to the goals of sustainable forest management are discussed.

Economic Concepts of Stumpage Values

The development of concepts associated with natural resource values has a long history in economics. Many of the ideas that have evolved in this vein involve concepts of economic rent. In the paragraphs that follow, we begin with the general theory of economic rent and then build on this concept to describe values of trees, land to grow trees, and the impact on tree and land values of regulations embodied in forest tenure systems.

Economic Rent

Economic rents are values of resources in situ (i.e., in the case of stumpage fees, trees standing in forests). However, it is difficult to undertand how natural resources in their raw form are valued without thinking about what they may become. The fact that standing trees can be used to make timber products gives them value.[4] The demand for standing timber is derived from the demand for logs, pulp chips, and other primary timber products. Demands for logs, chips, and so on are, in their turn, derived from the demand for secondary timber products such as lumber, board products, and wood pulp. Finally, the demand for secondary timber products is derived from demands for final products such as residential and non-residential construction and paper of various types and grades. To move up from one stage in the production chain to the next involves conversion expenditures and results in value being added to the product lower down the chain. Because values of standing trees emanate from the end products that they may become, value chains can be used to derive economic rents.

Economic rent can be defined as the return to a scarce input factor (in this case, standing timber) net of the total opportunity costs of bringing it into production.[5] The notion of economic rent can be most easily explained using a simple conceptual model (see Figure 5.1). In this figure, the pie represents the gross value of a forest stand in terms of the products that can be manufactured from the trees in the stand. These products might be logs or more highly manufactured products. The size of the pie represents the maximum gross value of the stand, assuming the wood is allocated to its most valuable use.

In this model, the costs are divided into the cost of materials – including energy; capital, comprising interest and depreciation; labour, including wages, salaries, and payroll costs; and profit – the return to entrepreneurship.[6] These costs represent all the outlays necessary to generate the gross value represented by the pie. If the gross value is measured in terms of logs, the costs are those necessary to harvest the timber, convert it into

Figure 5.1　Gross timber value and economic rent

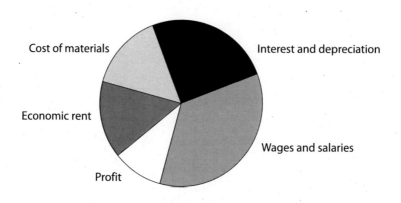

logs, and transport it to market. If the gross value is based on more highly manufactured products, the costs will also include manufacturing expenditures. Every stand of trees can have such a gross value assigned to it, but the costs and economic rents for each stand will be different.

If all costs are deducted from the gross value of the stand, there may be, as in this example, a surplus. That is, the value of the stand is greater than the total costs that are necessary to generate this value. This surplus is "economic rent," as seen in Figure 5.1, which is the return to the timber resource itself. Note that economic rent is a residual. It is the amount that remains after all costs of bringing the resource to market have been accounted for. If costs are just equal to the gross value of the stand, the economic rent is zero. Such a stand is said to be "marginal"; an operator harvesting it just breaks even by recovering all costs, including an acceptable, or normal, profit. If costs exceed the value of the stand, it is said to be "submarginal." If a submarginal stand is harvested, the return it generates will not cover all the costs of bringing the wood to the marketplace. This loss has to be absorbed by underpaying one or more of the productive inputs. Such losses are usually absorbed, initially, by reduced profits. A firm that consistently harvests submarginal timber cannot survive over the long run and will seek other opportunities for its capital.

The economic rent of standing timber is the potential maximum return to the owner of the timber. It is, in fact, the maximum amount forestland owners, including governments, can collect while enabling those harvesting the timber to cover all their costs. However, how much of the economic rent is actually collected by the owner of timber depends on how stumpage rates are determined.

This discussion thus far concentrates on concepts associated with standing trees that are assumed to already exist. However, in the context of forest management, we are also interested in growing trees. To accommodate the prospect of growing trees within the concept of economic rent, it becomes useful to distinguish situations where trees must be grown first before they are harvested.

Stock Rents and Forestland Rents

Since Crown forest tenure holders in Canada are frequently in control of both timber-harvesting and timber-growing operations, governments may be concerned with collecting economic rents from two sources of value. First, there are mature stocks of timber, which may be used to produce value-added products. The values derived from this source, which are the focus of the above discussion, may be referred to as "stock rents." However, as mature stocks of timber are depleted, the forestland that remains may be used to produce future timber crops. Such values are often enhanced by investments in silviculture, including reforestation and stand management. If forestlands are sufficiently productive to produce benefits greater than the costs associated with growing new stands, then surplus values, that can be referred to as "forestland rents," may be generated.

Luckert and Haley (1990) have shown how the method of rent collection is crucial in determining whether forest companies have incentives to voluntarily invest in silviculture. Stumpage fee systems based on timber harvesting alone – stock rents – will, if used to collect forestland rents, consistently overestimate the rent available, since they fail to take account of the additional costs associated with producing new forest crops. If stumpage fee systems for cultivated timber crops do not recognize these extra costs assumed by tenure holders, incentives for investing in silviculture are eroded, and the amount of available forestland rent is reduced.

Social Rents

Thus far we have incorporated the various costs of growing and harvesting timber into the determination of economic rent. However, as discussed in Chapter 2, considering resource values within the context of property rights involves acknowledging the presence of social conditions that may also influence resource values. Such considerations take us into the realm of "social rents," which include the benefits the Crown receives from imposing costs of requirements on tenure holders.

The actual benefit streams available to licensees from timber harvesting, which comprise their property rights (see Chapter 2), may be significantly reduced by the costs associated with the various conditions that must be met in order to fulfill their contractual obligations. In this context, the total returns collected by governments from licensees for their use of public lands can be regarded as having two components: direct cash returns, principally in the form of stumpage fees, and returns in the form of social benefits – recreational and environmental services, for example – that are the result of the obligations that licensees must fulfill in order to maintain their access to public timber. The greater the obligations placed on tenure holders, the less cash will be available for the Crown to collect. This non-cash component of governments' returns has been referred to as social rent (Luckert 1991a). Social rents are collected if the benefits society receives from regulating tenure holders exceed the costs borne by them (compliance costs) as a result of their rights having been attenuated, plus the costs to governments of enforcing the regulations (enforcement costs). If the total costs of regulation exceed gains in social welfare, rent is being dissipated through the use of inefficient tenure policies.

Analytical Framework for Comparing Provincial Stumpage Fee Systems in Canada

Because of the importance of stumpage fees, governments have expended significant efforts in their design. As is the case of tenures and forest practices, the complexity of stumpage fee systems calls for the identification of a number of salient characteristics that are common to all systems and allow them to be compared. Moreover, just as tenures and forest practices were described in previous chapters in terms of those features that are thought to have important forest policy implications, the framework that follows is based on characteristics of stumpage fee systems that may influence their efficacy in achieving public goals. It includes the general methods used to determine stumpage fees; the way in which stumpage fees are levied; how stumpage revenues are distributed; the extent to which stumpage fee systems reflect variations in species, wood quality, intended end use, and location; and how stumpage fees are adjusted in response to changes in market prices, costs, and other variables.

For many provinces, there is only one stumpage fee system. However, for some there may be multiple stumpage fee systems in place. For the purposes of this book, with the exception of British Columbia, we concentrate on the primary stumpage fee system in each province. Each of these systems is described according to the following characteristics.

Types of Stumpage Fee Systems
The type of stumpage fee system refers to the general structure of the arrangements used to determine stumpage fees. Within each province, the stumpage fee system(s) may be considered to belong to one, or a combination of, four general categories.

Fee Schedules
Essentially, under this system, a fixed charge is levied per cubic metre of wood, which is usually predetermined when a tenure is awarded. The charge may be varied by species, product category (e.g., pulpwood, saw logs), or geographic location, but these refinements do not change the basic attributes of the system. Schedules may be set unilaterally by governments or negotiated individually with each tenure holder. They may take the form of a schedule attached to a statute, a regulation, or be part of a licence document.

Appraisals
Instead of setting specific values of stumpage fees in schedules, some stumpage fee systems specify processes that will be followed for appraising stumpage values. This approach uses various techniques to estimate the actual value, or economic rent, of a forest stand. Appraisals

take two main forms – market appraisals and engineering/accounting appraisals.

Market appraisals can be used where reliable competitive market data for standing timber sales exist. This approach is very much like real estate appraisal. Recent market prices for timber with similar characteristics to the parcel being appraised are collected and their average price adjusted to take account of any special characteristics of the timber in question.

Engineering/accounting appraisals are based on market prices for logs, lumber, or other further manufactured products and the costs of producing these products as estimated from engineering (for example, machine types and productivities, labour productivity) and accounting information. Product prices minus production costs, including an allowance for entrepreneurial profit, yield an estimate of economic rent, which can then be used to set stumpage price.

Competitive Auctions

This approach requires that timber be advertised, often at a reserve or upset price, and either sold to the highest bidder above the reserve price or to the bidder that meets both pecuniary and other criteria, such as an undertaking to build and operate a certain type of manufacturing plant or create a certain number of jobs.

Hybrid Systems

Stumpage fee systems may use combinations of the aforementioned system types. For example, a minimal reserve price for competitive auctions may be determined by appraisal, or the government may administratively set province-wide or regional fee schedules based on appraisals.

Method of Payment (Fixed or Variable Costs)

Stumpage fees may be collected on a volume basis or on an area basis. If collected on a volume basis, the buyer must pay a fee for each cubic metre of wood harvested. If the system is area based, a fee is levied per hectare made available for harvesting. As far as a buyer is concerned, volume-based fees comprise a variable cost that fluctuates with the volume of wood harvested, whereas area-based fees are a fixed cost per hectare regardless of how much wood is removed.

Distribution of Stumpage Fees

Stumpage fees that are collected by provincial governments may be distributed in a number of ways, to be used for differing purposes. Three general alternatives regarding where stumpage fees go are recognized:

> *general revenue* – to be used for whatever purposes the provincial government deems appropriate;
>
> *provincial forest ministry or department* – to be used for forest administration, management, and regulation; or
>
> *dedicated forest management/research funds* – which may be used by either private firms or public agencies for designated purposes.

In some cases, the distribution of stumpage revenues is split between two, or possibly all three, of these target areas.

Resolution in Assessing Fees

Provincial stumpage fee systems display differing degrees of complexity in their degree of resolution in assessing stumpage fees. At one extreme, there may be a single schedule of stumpage fees for all species over the entire province. At the other extreme, stumpage fees may vary depending on the species, wood quality, and costs associated with a specific harvesting block. Three major dimensions over which variability can occur are recognized:

> *product variability* – stumpage fee assessments take account of the end products for which the timber is to be used – for example, lumber or pulp;
>
> *species variability* – stumpage fee assessments take account of different species and/or species groups; or
>
> *location variability* – stumpage fees are assessed separately for each geographical region, tenure, or harvesting block.

In cases where the resource base is fairly homogenous, resource values may show little variability, and little resolution in assessing fees may be necessary, thereby saving administrative costs. However, where multiple species, multiple products, and varying harvesting and transportation costs are significant factors, the extra costs associated with a higher resolution system may be warranted.

Adjustments for Changing Market Conditions

As prices of forest products and operating costs experienced by forest companies change, so do standing timber values. Therefore, once base levels of stumpage fees are set, many provinces adjust these amounts over time, thus avoiding frequent reappraisals. Variants on how these adjustments are made are characterized for this analysis as follows:

> *product prices and/or input costs* – Are adjustments to stumpage fees based on changes in product prices and/or input costs?
>
> *frequency of adjustments* – How often are stumpage fees adjusted? What reference period of changing prices is used to adjust stumpage fees? How soon after the reference period do stumpage fee changes take effect?
>
> *method of adjustment* – What are the scalar/formulas used to adjust base stumpage fee levels for changes in market prices?
>
> *maxima/minima* – Are there maximum or minimum constraints on stumpage prices as prescribed or calculated?

Reviewing and Revising Stumpage Fee Systems

As discussed in the previous sections, many stumpage fee systems have arrangements to provide for revisions in response to changes in product prices and costs. However, there are numerous additional factors that may bring about changes in stumpage values over time. For example, new technology may render species that were previously classified as "weed species" marketable, or changes in licensees' obligations under their tenure contracts may increase their operating costs. Given such eventualities, all the elements of stumpage fee systems that affect prices may be subject to periodic review and revision.

Amount of Stumpage Fees Paid

A comparison of stumpage fees between regions is a dangerous game. Different resource characteristics, market structures, distances from markets, and property rights to timber make useful comparisons of collected amounts relative to economic rent availability difficult. Nonetheless, the range of stumpage fees collected discloses information about how large a price signal stumpage fees represent in alternative jurisdictions.

> *Amounts of fees charged* – How much is generally charged for harvesting wood? The number of different fees charged will depend on the resolution of the specific stumpage fee system.
>
> *Special stumpage fee schedules* – Under some circumstances (e.g., for salvage after fires or insect infestations), special stumpage fee schedules may be used. How much is charged under these circumstances?
>
> *Timing of payments* – How long after harvesting are stumpage fees due?

Interprovincial Comparison of Stumpage Fee Systems

In this section, we summarize information about stumpage fee systems in a form that allows interprovincial comparisons to be made. Table 5.1 summarizes the characteristics of stumpage fee systems by province.

Types of Stumpage Fee Systems

For the most part, provinces use one stumpage fee system for all tenure types. Exceptions can be found in British Columbia and Alberta, where both competitive auctions and appraisals are used. These two provinces hold the only competitive auctions of public timber in Canada. In Alberta, auctions are restricted to timber sales provided for small operators, whereas in British Columbia, auctions for short-term timber sales licences, administered by BC Timber Sale Licenses, are open to all bidders. Auctions play a small role in Alberta, but their role in British Columbia has been increasing, motivated at least partially by concerns over US countervailing duties on softwood lumber. Today, more than

Table 5.1 Interprovincial comparison of stumpage fee systems

Province	System type	Fixed versus variable costs	Distribution of fees	Resolution	Market adjustments	Review and revision
British Columbia	Appraisals and auctions*	Variable	General revenues	Product, species groups, and area	Product price adjustments; quarterly	Annual
Alberta	Hybrid: schedules and appraisals; and auctions*	Variable and fixed*	General revenues and dedicated funds	Product, species groups, and area	Product price adjustments; monthly	In 2008
Saskatchewan	Schedules	Variable	General revenues	Product, species groups, and tenure (FMAs)	Product price adjustments; quarterly	No set time
Manitoba	Hybrid: schedules and appraisals	Variable	General revenues	Product, species groups, and area	Product price adjustments; monthly	No set time
Ontario	Hybrid: schedules and appraisals	Variable	General revenues and dedicated funds	Product, species groups, and area	Product price adjustments; monthly	Annual

Quebec	Schedules	Variable	General revenues and reimbursements	Product, species groups, and area	Product price adjustments; quarterly	Annual
New Brunswick	Hybrid: schedules and appraisals	Variable	General revenues	Product, species groups, and area	Product price adjustments; annually	Annual
Nova Scotia	Hybrid: schedules and appraisals	Variable	General revenues and dedicated funds	Product, species groups, and area	Product price adjustments; annually	Every 5 to 10 years
Newfoundland and Labrador	Schedules	Variable	General revenues	Product and road access	No automatic adjustments	Annual

* Vary by region and/or type of tenure.

20 percent of British Columbia's annual harvest from Crown lands is sold competitively. On the BC coast, the price information revealed by these sales is used as a starting point for a market-based appraisal system for timber held under long-term tenure arrangements. In the BC Interior, stumpage fee determinations for long-term tenures rely on the comparative value timber pricing system – an engineering/accounting appraisal method.

The most common stumpage fee systems in Canada are hybrids incorporating schedules and appraisals. Exceptions are in Saskatchewan and in Newfoundland and Labrador, where systems rely solely on schedules. British Columbia is the only province that does not make use of schedules, relying entirely on appraisals and competitive auctions.

Method of Payment (Fixed or Variable Costs)
Almost all stumpage payments in Canada are structured as variable costs, usually assessed on a log scale (m^3) basis. In Alberta, small volumes on commercial timber permits are sold using lump sum sales, assessed on a per-hectare basis (fixed cost). In British Columbia, the Forest Act provides for the use of area-based sales, but this provision is rarely invoked.

Distribution of Stumpage Fees
In all provinces, stumpage fees go, at least in part, into general provincial revenues. However, in some cases a portion of stumpage fees is diverted into special funds dedicated to specific forestry objectives – for example, for forest renewal or research into forest management. Provinces with these types of dedicated funds are Alberta, Ontario, and Nova Scotia. In Quebec, there is not a specific fund within which stumpage fees are collected; however, stumpage fee payments are reduced according to credits that tenure holders can earn for undertaking approved silvicultural operations.

Resolution in Assessing Fees
Across Canada, there is a significant degree of resolution in stumpage fee systems. Most provinces differentiate stumpage fees based on endproducts, species, and area. All provinces differentiate stumpage fees, to

some extent, by product. Stumpage prices may be differentiated according to prices of minimally value-added products, such as logs, whereas in other cases, differentiation may be based on further value-added products, such as oriented strand board in Alberta. All provinces except Newfoundland and Labrador also differentiate stumpage fees according to species. However, in British Columbia, Ontario, and Quebec, fees are differentiated by specific species, whereas in the remaining provinces (save Newfoundland and Labrador), fees are differentiated by species group (generally coniferous and deciduous).

All provinces also differentiate their stumpage fees according to some type of area-related variable. In most provinces, these distinctions arise as schedules that apply in different zones. However, in Saskatchewan and British Columbia, fees are differentiated by tenure. In Saskatchewan, each forest management agreement has a specific stumpage fee schedule. In British Columbia, holders of community forest licences pay stumpages at reduced rates. In Newfoundland and Labrador, fees are differentiated by whether or not there is serviced road access.

Generally, the more heterogeneous the timber supply and the more valuable the timber resource, the more discriminating the schedule of stumpage fees.

Adjustments for Changing Market Conditions

All provinces have mechanisms in place to adjust stumpage fees to changing market conditions. In most provinces, these adjustments take place systematically according to established formulae based on forest product price indices derived from various sources – Statistics Canada or trade journals such as *Madison's Lumber Reporter* and *Random Lengths*. The one exception is Newfoundland and Labrador, where stumpage fees are not subject to regular, formulaic adjustments but are reviewed periodically by cabinet. In the framework used here, this is considered to be a revision rather than an adjustment and is considered in the next section. None of the market adjustments in the provinces that have automatic formulaic adjustments include changes to input costs. Established formulae are used to adjust stumpage fees on a monthly, quarterly, or annual basis.

Reviewing and Revising Stumpage Fee Systems

Most stumpage fee systems are reviewed and revised on an annual basis. The Prairie provinces provide exceptions; Saskatchewan and Manitoba have no set revision schedules, and Alberta's system was recently reviewed and revised. Periodic revisions in Nova Scotia vary by tenure in the range of every five to ten years.

Amount of Stumpage Fees Paid

Amounts charged for stumpage fees are highly variable between and within provinces. Differences in end products, species, and location cause stumpage fee formulae to create large and varied ranges of fees. The different ways that provinces report average stumpage fees can also affect these ranges. Some ranges reflect values averaged over large areas and conditions of the province, whereas some averages are reflective of a specific product-species-zone only. Because of the large variation in these values, we do not attempt to summarize them in Table 5.1.

Desirable Attributes of a Public Stumpage Fee System

In assessing public stumpage fee systems, it is useful to do so in the context of general characteristics that could be considered desirable attributes of such schemes. In this section, we present these attributes and then, in a subsequent section, relate them to the goals of sustainable forest management. These attributes can be classified under four headings: efficiency, equity, flexibility, and simplicity.

Efficiency

Economic efficiency is frequently concerned with net benefits – that is, benefits less costs – accruing to a society as a result of an economic activity or public policy, in this case, the sale of public timber. Economic efficiency is typically concerned with the aggregate well-being of society; it is not concerned with how the benefits and costs of an economic activity are distributed among a society's members.

To be efficient, a stumpage fee system should ensure that when a stand of timber is harvested, the maximum net value of the timber – that is, its economic rent – is generated. Furthermore, the system should not distort the resource allocation decisions of timber buyers in ways that are contrary to the public good. For example, a system that generates stumpage prices that are less than the full worth of the stands being sold may create inefficiencies by sending market signals that discourage firms from devoting resources to processing technologies that conserve wood. In Canada, because the majority of firms manufacturing forest products purchase a high proportion of their wood from provincial governments, such distortions can dictate the efficiency of an entire forest industry. Finally, the costs of determining stumpage fees and collecting stumpage revenues should be at the minimum level consistent with the overall objectives of public policy.

Equity

Whereas efficiency is concerned with the net benefits of an economic activity or public policy to the economy as a whole, equity, or fairness, is concerned with the distribution, or incidence, of benefits and costs. In the context of this discussion, questions of equity involve who bears the costs and who benefits from public stumpage fee systems, and whether this distribution meets societal standards of fairness.

If society believes that the owner of the resource should be entitled to the economic rent, then an equitable stumpage fee system should capture the full, available economic rent from timber resources on behalf of the public owners of Crown resources. In contrast, an inequitable system will either leave windfall returns for timber buyers or for other factors of production (e.g., labour) that may capture more than their market values as inputs into production.

An equitable stumpage fee system also requires that all buyers be treated in an even-handed way by providing a level playing field for competition. For example, two parcels of timber that are in every way similar should be put on the market at the same price. Conversely, two parcels of timber that differ in ways that influence their values should be put on the market at different prices reflecting their differences in value. In short, an equitable stumpage fee system would afford each firm the same opportunity to be as profitable as the next.

Flexibility

Flexibility requires that stumpage fee systems can be adjusted, or can self-adjust, in ways that reflect the dynamic nature of forest products markets and buyers' operating costs. Input costs and wood products prices can change markedly, and in many cases rapidly, over time. These changes cause values, or economic rents, of forest stands to change both cyclically and secularly. For example, increasing product prices cause economic rents to rise, whereas increasing input costs cause rents to fall. If stumpage fee systems are not sufficiently flexible to accommodate these changes, they will fail to effectively estimate economic rents, and both efficiency and equity may be compromised.

Simplicity

Despite the complexity of the issues that stumpage fees systems seek to address, simple and clear solutions may be preferable to more complex approaches. Basically, the system should be as simple as possible commensurate with other public objectives and constraints.

Complex approaches tend to be costly for governments to administer and for firms to follow. They may result in spurious precision if the framework for calculating stumpage fees is more complex than the input data can support. Simple systems also tend to be more transparent and more easily understood by both buyers and sellers. Transparency usually results in a more equitable system and reduces uncertainty on the part of licensees.

Implications of Crown Stumpage Fee Systems in Canada for the Pursuit of Sustainable Forest Management

The above discussion reveals a number of issues that may have significant impacts on the ability of provinces to pursue sustainable forest management. In the following paragraphs, we begin with issues associated with the failure to distinguish between stock rents and forestland rents and the failure to recognize social rents. We then discuss various issues following the stumpage fee characterization framework presented above.

Stock Rents versus Forestland Rents

As discussed earlier, the economic rent collected from harvesting primary timber crops – stock rent – is different from forestland rent generated by second-growth stands that have resulted from the investment activities of forest owners or licensees. Stock rent is the gross value of a forest stand net of all harvesting costs from standing tree to market. In calculating forestland rent, costs must include the amount invested in the production of a new timber crop and consider that harvests from newly established crops will not occur for many years. Regeneration and other silvicultural costs may be nominal if a stand is the result of natural regeneration and there are minimal expenditures on stand management; however, if the stand has been planted and/or intensively managed, silvicultural costs might be considerable and may significantly reduce the amount of rent available for collection as stumpage fees. Moreover, discounting future benefits of harvests far into the future can also seriously reduce rents.

No provincial government in Canada distinguishes between stock rents and forestland rents when collecting stumpage fees. Primary timber stands and those resulting from silvicultural investments are treated identically under provincial stumpage fee systems; consequently, incentives for more intensive forest management are weakened. A licensee that invests in timber production knows that there will likely be no return on this investment; it will be collected as stumpage fees when the crop is harvested. As a result, there may be little incentive for Crown forest tenure holders to voluntarily ensure timber crops are re-established after harvesting or to invest in the management of immature forest stands.

Given the weak incentives for voluntary expenditures on reforestation, all provinces require most tenure holders to reforest after harvesting (see Chapter 4). Thus, licensees regard expenditures on reforestation not as investments in future timber crops but as a current cost of doing business that is necessary to retain their rights to harvest Crown timber. Under these circumstances, reforestation is not optimal, in the sense of getting the greatest return for the dollars spent, but is simply carried out at the lowest outlay necessary to meet the regulations (Luckert and Haley 1990). In most cases, forest tenure requirements in Canada do not include expenditures on silviculture beyond reforestation. Thus, in the absence of either obligations or incentives, few stands are intensively managed.

Another factor that may prevent intensive management in Canada is the inherent low productivity of much of the forestland base. Even if tenure systems provided incentives for firms to consider future benefits and costs of management, the financial returns to such actions on many lands would be too low to justify expenditures (see, for example, Rodriguez et al. 1998).

Failure to Recognize Social Rents

As discussed earlier in this chapter, the economic rent collected by provincial governments from the operation of Crown forest tenures may consist of two components: pecuniary rents, which are collected as stumpage fees, and social rents. Governments collect social rents if the benefits, in the form of social goods and services such as recreation, biodiversity, and wildlife habitat, received from regulating tenure holders' activities exceed the costs imposed on them through the attenuation of their rights.

Recognizing that the total rent generated by forest tenures includes both pecuniary and social components presents governments with an entirely new set of costs to consider in determining stumpage values. Although research has resulted in empirical evidence of such costs, integrating such charges directly into stumpage fee calculations is difficult given the wide variety of restrictions placed on licensees and the difficulties inherent in determining compliance costs.[7]

If the costs of generating social rents are not adequately taken into account when determining stumpage fees, the stumpage fees collected may significantly reduce the value of tenure holders' rights to the point where they are no longer earning a normal profit. Under these circumstances, forest sector operations may be rendered unsustainable over the long run.

Although no provincial government explicitly considers social rents as elements of stumpage fee appraisal calculations, the social values generated by forest tenures may be implicitly recognized by charging stumpage fees that are below market value. For example, in British Columbia, community forests are explicitly treated differently, with holders of that tenure paying only a relatively small percentage of the stumpage rate paid by other tenure holders. However, for large tenure holders, allegations of below-market stumpage fees have been the major bone of contention in the long-standing softwood lumber trade dispute between Canada and the United States. Consequently, in recent years there has been pressure on the Canadian provinces to charge stumpage fees that better reflect the value of wood in forest products markets. Thus, the ability of the provinces to recognize the social values generated by forest tenures by means of adjustments to the amount of stumpage fees collected has become progressively limited. This problem is compounded by the fact that tenure holders' responsibilities have increased considerably as the demand for non-timber forest values, many of them in the form of public goods, has burgeoned.[8] For example, as we have seen in Chapter 4, forest practices regulations have proliferated across the country in recent years, imposing high compliance costs on tenure holders largely without commensurate adjustments in stumpage fees in recognition that these costs are generating significant net social benefits.

Types of Stumpage Fee Systems

Fee Schedules
In Canada, most provinces' stumpage rates are set out in a schedule. Stumpage rates may be simply set periodically by government or, as in most cases, be determined through some form of appraisal or negotiation. How these schedules are set may have serious consequences for sustainable forest management.

A crucial aspect of setting fees is transparency, which is served by accessible, understandable, and predictable procedures. However, a number of circumstances may obfuscate the determination of fee schedules. For example, closed-door negotiations between tenure holders and provincial governments may create the perception of an uneven playing field among tenure holders. If there is little confidence in the methods used to set fees, or if governments can adjust schedules at will with no recourse to established procedures, as is the case in some provinces, then stumpage fee systems incorporating scheduled rates can create a great deal of uncertainty among timber buyers. This insecurity may have undesirable consequences for forest sector investment.

Despite these potential problems, there are key benefits of scheduled stumpage fees. First, such approaches are potentially simple to understand and administer. In cases where resource values are low and/or uniform, such simple approaches may be warranted. Moreover, negotiations that establish such schedules may be sufficiently flexible to explicitly consider differing costs considerations of tenure holders and costs of tenure obligations.

Appraisals

Market appraisals, such as those used on the BC coast and in New Brunswick, can, if based on reliable data from competitive markets, provide estimates of economic rent that are efficient, equitable, and transparent. However, like any appraisal, whether it is for timber, real estate, antiques, or rare coins, the process calls for a fair amount of subjective judgment and results rely heavily on the skill and experience of the appraisers. Also, markets may be imperfect, with either some large buyers or large sellers able to control prices. This problem is commonplace in timber markets.

Problems also arise if timber sold in the markets upon which appraised prices are based is non-representative, in terms of species, age, condition, and location, of the timber being appraised. Adjustments can be made to account for such discrepancies, but such adjustments introduce elements of complexity and subjectivity to the process.

Finally, where market appraisals are used to determine stumpage fees to be charged for timber harvested from long-term Crown forest tenures – in British Columbia, for example – a further problem arises. As we have seen, forest tenure holders face many costs associated with meeting the obligations laid out under the terms of their tenure contracts. Consequently, in order to arrive at equitable stumpage fee estimates, it is necessary to estimate these costs and adjust the market data accordingly. This is a task fraught with difficulties and may further undermine tenure holders' confidence in the system.

In theory, stumpage fee appraisals based on engineering/accounting data can be both efficient and equitable. In practice, however, the method may be far from satisfactory. The main difficulties surround cost estimates. To begin with, it is not generally in tenure holders' interests to divulge their costs, so reliable information is difficult to come by. Moreover, the appraiser has to cost out every phase of the operation before it takes place – a task that is inevitably very complex. Logging costs are a function of a broad array of tree, stand, and site variables. If appraisals are based on market prices of further manufactured products, manufacturing costs must also be appraised, thus introducing further complexity.

A major problem in estimating costs is to make an allowance for entrepreneurial profit. Consequently, appraisers tend to ignore this element of cost completely or resort to subjective rules of thumb. For example, when this type of appraisal – the Rothery method – was used in British Columbia prior to 1987, a somewhat arbitrary 6 percent of operating costs was allowed for profit on the coast and 10 percent for the Interior.

Further problems occur if appraisals are based on end-product prices. As discussed in Chapter 3, most provinces require that firms holding long-term forest licences own and/or operate a wood products manufacturing plant. The absence of log markets can therefore create the need for appraisals to work backwards, from prices of processed forest products rather than from logs. This situation can create numerous problems.

First, the farther down the value-added chain a price is used, the more complicated are the appraisal calculations that subtract costs associated with each value-added stage.

Second, even if appurtenancy requirements allow for trades of logs between mills, stumpage fee systems based on end-product prices may prevent it. For example, a stumpage fee system may charge a tenure holder $25/m³ for a log destined for an oriented strand board plant, whereas another tenure holder may be charged $5/m³ for an identical log destined for a pulp mill. Under these conditions, a log could be worth much more in the oriented strand board plant but nonetheless end up in the pulp mill because of the lower stumpage fee associated with this allocation. With such a stumpage fee structure, a tenure holder has limited incentives to develop log allocation strategies that add the most value to the resource and maximize provincial stumpage revenues. Moreover, incentives to discover new uses for wood fibre are reduced, as stumpage values differentiated by end products tend to perpetuate wood flows into current uses. Such systems, by preventing the allocation of logs to those end uses and mills where the greatest net revenues can be generated, may result in reduced economic rents available for collection through stumpage fees.

Competitive Auctions

The problems associated with stumpage fee schedules and appraisal approaches to estimating economic rents can be largely overcome by a well-functioning system of competitive auctions. This is the method used for public timber in the United States and in several European countries (Haley 2006) but, as we have seen, in Canada only British Columbia sells a significant proportion of its annual harvest competitively.

Competitive auctions encourage each buyer to be as cost-efficient as possible so that it may be in a position to place the winning bid. Thus, there are strong incentives to put the timber to its highest value use and minimize costs. Consequently, maximum economic rents tend to be generated for the timber seller. Risk is borne by buyers that bid what they believe the timber is worth to them. Bids reflect the value of timber to the buyer net of all costs, including the costs of meeting tenure obligations, and are sensitive to product market and cost trends. Such a system can be efficient, equitable, and relatively inexpensive to administer. It is transparent and easily understood by all participants. In short, competitive markets may meet all the criteria of a desirable stumpage fee system.

But not all stumpage fee systems based on competitive auctions effectively promote efficiency and equity in timber pricing. For competitive auctions to work most successfully, buyers and sellers must have access to adequate information about the market, and the number of buyers and sellers must be sufficient to ensure that neither can influence prices. These conditions are frequently violated in timber markets. Where the principal seller is a large public agency, as in most Canadian provinces, it enjoys monopolistic powers that it can exercise to control market supplies of timber and, hence, prices. This inescapable fact undermines the confidence of buyers in market processes, since there are no guarantees that the market transactions will produce equitable prices. In Canada, it is frequently argued that buyers (generally large integrated forest companies) and sellers (generally provincial governments) may both exert market power. Such a market structure is known as bilateral monopoly.[9] In these circumstances, contractual stumpage prices are negotiated between buyers' and sellers' representatives. Actual price levels, which determine the distribution of economic rent, will depend on the relative bargaining strengths and skills of the parties involved (Nautiyal 1980).

Another common problem occurs in markets when there is one, or a small number, of large buyers and a large number of small sellers.[10] In such circumstances, buyers are in a position to set prices and capture much of the economic rent that equity considerations suggest should accrue to the forest owners. This is often the case in regions where there is a large number of small private timber producers supplying a small number of large manufacturing companies. However, it may also occur where there is a relatively large number of small Crown forest tenures, as is the case in British Columbia, where, in some regions, woodlot licences, First Nations forest licences, and community forest agreements supply a growing proportion of stumpage markets. In some cases where such conditions exist, forest owners and/or licensees organize into cooperatives in order to present a united front to the buyer or buyers and, as described above, create bilateral monopolies.

Another problem associated with competitive auctions in Canada is asymmetry among different classes of buyers. This may occur in two forms. First, if large corporations and small mills and market loggers compete in the same market, the large companies that have access to substantial capital reserves may be willing to take losses on timber sales over the short term by bidding more than the timber is worth and driving out their smaller competitors, thus reducing future competition. Second, holders of different tenure types competing in the same marketplace for public timber may face very different costs in relation to their contractual obligations under their agreements, thus placing those who bear higher costs of providing public services at a competitive disadvantage. A solution to both these sources of asymmetry might be to segregate markets – for example, into small companies and large companies or into different classes of tenure holder – in order to level the playing field.[11] Such segregation, however, can present administrative problems and reduce the number of bidders in many markets to the point where the market has too few participants to be considered competitive.

Hybrid Systems

Hybrid systems may borrow strong points from each individual approach. For example, if competitive auctions are feared to be noncompetitive, the government can protect itself from low bids by using appraisals to establish an upset price (i.e., the price at which bidding must start). Moreover, these upset prices can be adjusted through negotiations between tenure holders and the government to reflect cost considerations that the government may not be aware of.

Method of Payments (Fixed or Variable Costs)

The provinces are similar in their methods of payment. With only minor exceptions, stumpage fees are paid on each cubic metre of wood harvested and are regarded by the buyers as a variable cost. This practice reduces the need for detailed cruises of standing timber and allows public forest managers to keep track, for yield control purposes, of the actual volumes harvested. However, unlike stumpage fees charged on a per-hectare or fixed-cost basis, this method of collecting stumpage payments modifies the behaviour of firms in terms of the volume of timber recovered. Because firms pay for each log harvested, they have an incentive to leave on the ground those logs of too low value to cover costs, including variable stumpage charges (Nautiyal and Love 1971; Vincent 1990, 1993; Hyde and Sedjo 1992; K. van Kooten 2002).[12] Foresters often refer to this as "high grading" the timber.

To overcome this problem, all provinces have regulations that impose utilization standards on harvesters that specify which trees are to be felled and which logs must be harvested. Failure to remove logs that meet these standards may attract substantial penalties. However, utilization standards, which are usually applied over very large areas, are blunt instruments for ensuring that the volume recovered from each logging site is optimal from an economic perspective. In many cases, they create inefficiencies by reducing the economic rent that stands of timber can generate by forcing firms to harvest wood that costs more to recover than it is worth (Uhler and Morrison 1986).

It might be argued that a stumpage fee system that encourages more logs to be left on the ground is desirable in that unharvested course woody debris may contribute to sustainable forest management by facilitating greater ecosystem diversity. However, to achieve environmental policy objectives by reducing economic efficiency in a broad and unpredictable manner is bound to result in suboptimal solutions from both economic and environmental perspectives. Policies designed to promote sustainable ecosystems should be the subject of specific instruments designed to achieve desired outcomes as effectively and efficiently as possible.

Distribution of Stumpage Fees

Across Canada, the revenues generated by stumpage fees largely flow into general revenues, thereby providing governments with the flexibility to decide where public resource rents will generate maximum social benefits. However, several provinces have provisions allocating a certain proportion of stumpage revenues collected to forest sector activities such as forest renewal, silviculture, protection, and research.

From a sustainable development perspective, whether forest resource rents should remain in the forestry sector is an involved and contentious issue (Luckert and Williamson 2005). Given that sustainable development transcends the boundaries of forestry, it could be logical for governments to take rents generated in the forestry sector and use them to invest in the creation of capital elsewhere in the economy. Directing resource rents into general revenues is in accordance with the theory of weak sustainability, which holds that it is not important to maintain all (e.g., forest) capital stocks at a constant level. Rather, it is the aggregate capital stock that a society needs to maintain. Therefore, drawing down one stock (e.g., forests) is acceptable if another stock (e.g., hydroelectric capacity) increases. This approach to sustainability maintains that sustainable development depends on society creating and maintaining an optimal mix of capital assets. Indeed, most provinces in Canada have historically used wealth from forests to facilitate general economic development, and today, for several provinces – including British Columbia, Quebec, and New Brunswick – stumpage revenues remain a significant proportion of provincial budgets.

However, a contrary view – that of strong sustainability – suggests that capital is not substitutable and that it is, therefore, important to sustain individual capital stocks. That is, there may be something unique about forests that implies we should maintain them in their current state in perpetuity, rather than allowing the wealth they are capable of generating to be captured and transferred to other sectors of the economy. In particular, forest resources such as biodiversity and endangered species may be in specific need of strong sustainability constraints. In practice, a compromise between these two approaches is likely called for. Wealth transfers between sectors of the economy are necessary, while ensuring that irreversible damage to unique capital stocks is avoided. The practice of distributing stumpage fees between general revenues and dedicated forestry funds seems to reflect this compromise.

Resolution in Assessing Fees

Schedules generally discriminate by species or species groups and location, but may also incorporate the end product for which the timber is destined to be used, and some indicator of relative harvesting costs. How much stumpage fee systems discriminate is largely driven by the degree of heterogeneity of stumpage values versus the costs associated with capturing these differences with a stumpage fee system. The more homogenous the resource values and the more costly it is to have higher resolution, the less resolution a stumpage fee system should have. Conversely, if there is great heterogeneity in high-value timber, the costs of a stumpage fee system with a high degree of resolution can be justified. This observation is consistent with Pearse (1988), who suggests a positive correlation between resource values and the complexity of property rights that evolve.

Adjustments for Changing Market Conditions and Reviewing and Revising Stumpage Fees Paid

The flexibility of stumpage fees to change with changing market conditions may be crucial to the forest industry and to governments. Given that most forest tenures control the volume of timber that can be harvested annually or periodically, forestry firms may have little flexibility in adjusting their harvest levels to market conditions. Consequently, if stumpage fees are not appropriately adjusted for market conditions, the impacts on licensees may be severe. For example, allowable annual cut (AAC) regulations may require tenure holders to harvest more than they wish during periods of low demand. If stumpage fees are also not adjusted to reflect lower product prices, forestry firms may not survive the downturn. Conversely, tenure holders may be prevented from harvesting as much as they wish during periods of high demand. During these times, if stumpage fees are not adjusted to reflect higher product prices, licensees may receive windfalls at the expense of government coffers.

Over longer periods, the adjustments built into stumpage fee formulae are likely to require review and revision. For example, the historic increases in harvesting constraints placed on tenure holders have influenced the profitability of harvesting operations and the rent left over for governments to collect. Moreover, changes in technology can significantly alter the costs associated with timber harvesting, thereby necessitating adjustments to assumptions of underlying cost structures in the industry.

Amount of Stumpage Fees Paid

As mentioned above, comparing stumpage fees is a dangerous game. However, the amounts actually collected are important. If stumpage fees collect less than the full economic rent, the returns that buyers derive from timber harvesting are greater than their costs, including the normal profit necessary to keep them in business, and, in the case of state-owned timber, the public purse suffers. Instead of going to the public purse, rents may flow to higher profits for the tenure holder, or higher wages and salaries for labour. If stumpage fees collect more than the available economic rent, timber harvesting firms receive below-normal profits – a situation that cannot be sustained in the long run. For example, Pearse (2001) observed that investment capital has been leaving the BC coast likely as a result of stumpage fees that exceed economic rents after all costs, including the costs of meeting obligations under tenure agreements, have been taken into account.

Moreover, the level of stumpage fees can have important implications for the overall structure of the forest industry. With governments in Canada owning vast amounts of forestland, the stumpage fees that they set have large impacts on domestic log values, including those logs trading in private markets. These prices provide important signals for firms contemplating investments in growing and processing trees. The price for standing trees represents the return for those contemplating silvicultural investments, whether on private or public land. Therefore, higher stumpage prices yield higher silvicultural returns. Moreover, to forest processing plants, stumpage prices represent input costs that can influence which technology they invest in and adopt. If stumpage prices are low, incentives to invest in wood-saving technology are reduced. Conversely, if stumpage prices are high, incentives to invest in wood-saving technology are also high.

Conclusion

Stumpage fees are collected by provincial governments on all timber harvested from Crown forest tenures in Canada. They are the most important component of Crown tenure holders' fiscal obligations and comprise a major attenuation of their rights. How stumpage fees are determined and collected can have important impacts on how tenure holders behave and, consequently, has significant implications for sustainable forest management – particularly the sustainability of a robust and competitive forest industrial sector.

In the context of sustainable forest management, several issues surrounding stumpage fee systems are of critical importance. A key issue concerns the methods used to calculate stumpage fees. Most provinces use price schedules under which set charges are levied per cubic metre of wood harvested. Such systems frequently suffer from problems related to efficiency, equity, flexibility, and perhaps, to a lesser degree, simplicity. In theory, competitive stumpage markets display all the characteristics of a desirable stumpage fee system. Yet only in British Columbia is a substantial volume of Crown timber – over 20 percent of the provincial AAC – sold by auction in competitive markets. The use of competitive sales across Canada has likely been limited by frequent situations where market demand is dominated by a small number of large integrated forest products manufacturers that limits the competitiveness of sales. In the face of this problem, one approach would involve substantial restructuring of the forest industrial sector, as has been partially achieved in British Columbia. But many regions of Canada have a small number of large firms for good reason. Economies of scale in the manufacturing of many forest products are such that smaller firms would be unable to compete globally (see Chapter 3). However, if we were to consider stumpage fee systems in the sole context of selling logs, rather than selling logs to be used for specific end products, we could potentially design more efficient and flexible arrangements for determining stumpage fees. Tenure arrangements that encourage log markets could increase competition as new entrants would be able to access wood supply and have incentives to use purchased wood efficiently. Such markets would allow logs to gravitate toward their highest-value use. Current stumpage fee systems that tie fees to end products perpetuate rigid and inefficient log allocations and may negatively influence the efficiency of entire forest industries.

Another serious problem with current stumpage fee systems is that they create disincentives for tenure holders to voluntarily invest in reforestation and the silvicultural management of forest stands. By failing to distinguish between timber values and forestland values, current stumpage fee systems remove incentives to create or enhance future forests.

Finally, it is difficult yet important for stumpage fee systems to recognize that governments collect a portion of the rent generated by Crown forest tenures in the form of non-timber goods and environmental services that benefit society at large. These benefits are provided at considerable cost to tenure holders in order to meet their contractual obligations. No province explicitly recognizes these social rents in their stumpage appraisals, but failure to do so may undermine the ability of Crown forest tenure holders to remain internationally competitive and sustain their operations.

In Search of Forest Tenures for Sustainable Forest Management   6

Since before Confederation, Crown forest tenures have been a major instrument – some would maintain *the* major instrument – applied to the implementation of forest policy in Canada. To a large extent, the evolution of public forest policy in the Canadian provinces has been a story of Crown forest tenure arrangements being adapted to acknowledge changing public attitudes toward forests, meet new demands on forest resources, and recognize evolving domestic and global economic conditions.

Contemporary Canadian forest tenures generally have their roots in mid-twentieth century Crown forest policy reforms that introduced the concept of sustained yield of timber. The almost exclusive emphasis on timber harvesting rights, strict control over periodic or annual harvests to create an "even flow" of timber, terms that span the amortization periods of large pulp mills or integrated forest products complexes rather than the rotations of timber crops, mill appurtenancy requirements, and the regulation of log exports are all evocative of an era in which harvesting old-growth timber was seen to be the main driving force of economic development, and policy makers held to the notion that regional economic prosperity and stability is achievable by ensuring that an even flow of logs reaches designated local manufacturing facilities. The veneer of environmental regulations that has been added to these pervasive attributes of most provincial Crown forest tenures has frequently exacerbated the problems inherent in a system that in many ways has failed to adapt to the new domestic and global, social, and economic realities of the twenty-first century.

The evidence presented in Chapters 3, 4, and 5 suggests that provincial forest tenure systems have many common design features that not only fail to promote the multi-faceted objectives of sustainable forest management but may render them difficult, or even impossible, to achieve. These problems largely stem from the fact that sustainable forest management confronts policy makers with vexing trade-offs, and provincial governments have struggled, with limited success, to identify innovative solutions to address these trade-offs. For example, although public policies in the Canadian provinces over the past fifteen years have seen a shift away from economic concerns and toward greater emphasis on environmental protection, provincial governments have not developed new, innovative institutions and management techniques designed to achieve the multiple objectives of sustainable forest management. Instead, as discussed in Chapter 4, they have mainly chosen to augment existing tenure arrangements with regulations that constrain the actions of timber harvesters in recognition of the increasing need to acknowledge environmental values and the burgeoning demand for non-timber, often non-consumptive, forest products and services. These regulations have considerably increased the responsibilities that tenure holders must assume in order to exercise their rights. Consequently, burgeoning compliance costs have significantly increased the total costs of timber production (G.C. van Kooten 1994; Haley 1996; Clarke 1997), with serious consequences for economic and social sustainability as the international competitiveness of forest industries in many parts of the country has deteriorated.[1]

In short, with the passage of time, Crown forest tenure systems have become increasingly complex until today they comprise a collage of rights of varying vintage which, because they were designed at different times to serve diverse objectives, vary substantially in the rights they convey and obligations they exact (Pearse 1992). Such tenure policies have sometimes left governments, industry, and the public confused about who has rights and who bears the responsibilities for what aspects of forest management.

As a result of this pattern of evolution, there appear to be significant gaps between the new norms of sustainable forest management and what existing forest tenures actually deliver. In short, we find today variations of anachronistic forest tenures that frequently fail to encourage forest practices that result in forest management that is environmentally, socially, or economically sustainable.

In this concluding chapter, we begin with a discussion of the current state of forestry in Canada. In the next section, we review links between tenure systems and the goals of sustainable forest management. We then describe how the forest industry is facing some long-term trends that suggest necessary changes to their current operating environments. In light of these challenges, the subsequent section highlights some of the current problems with tenure systems and also considers potential future directions to alleviate these problems. A review of these problem areas reveals several principles for change that could guide the search for new tenure policies. But we also recognize several potential barriers to change. We then proceed to consider how they might be overcome.

Tenure Systems and the Goals of Sustainable Forest Management

A major focus of this book is to examine today's Crown forest systems in relation to the goals of sustainable forest management that since the early 1990s have been the overarching objectives of forest policy, federally and in all provincial jurisdictions. To what extent do provincial Crown forest tenures further the goals of sustainable forest management, and what constraints do they present to the achievement of these goals?

When considering tenure arrangements in relation to the objectives of sustainable forest management, it should be recognized that the sustainable management of forests is a complex and multi-faceted policy goal. As explained in the Introduction, sustainable forest management is generally regarded as having three dimensions: environmental sustainability, economic sustainability, and social sustainability. Environmental sustainability requires policies that will ensure the structures of forest ecosystems are maintained in forms that will allow desired combinations of forest goods and environmental services to be produced now and in the future. Policies designed to further economic sustainability should ensure that forests are managed in ways that sustain, and ideally enhance, their contributions to incomes and employment – local, provincial, and national – over the long term. And, finally, the social component of sustainable forest management is concerned with policies that lead to the equitable distribution of the wealth generated by forests among individuals and groups within society and, particularly, among regions. Social sustainability is also increasingly concerned with meeting the aspirations of Aboriginal peoples for economic development and increased control over activities within their territories.

In pursuit of sustainable forest management, governments must endeavour to achieve an optimal balance among these three imperatives. Consequently, trade-offs are necessary between different objectives. For example, a policy that tries to maintain timber manufacturing jobs in all forest-based communities will inevitably fail in light of changes taking place in dynamic domestic and global economies and the changing mix of goods and environmental services demanded of forestland. Likewise, maximizing the volume of merchantable timber that can be harvested on a sustainable basis – a cornerstone of public forest policy in all provinces from the 1950s through to the 1990s – inevitably conflicts with the preservation of intact, sustainable natural ecosystems. Moreover, such conflicts tend to pit rural and urban populations against one another, as forest-based employment tends to be rurally based, whereas passive-use values associated with environmental protection and preservation tend to be urban-based.

These trade-offs are pervasive in forest tenure policies – so much so that it is hard to envision any policies that meet the simple Pareto improvement test, where at least one person is better off from a change but none is worse off. There are no doubt cases where a potential Pareto improvement – or increase in net social benefits – is possible, so that the welfare gained from those benefiting from policy change exceeds the losses. Even in this case, however, the inevitable trade-offs in tenure policy change pose formidable political obstacles to reform. We return to this theme in the section on barriers to change below.

Current Challenges Facing the Forest Products Industry in Canada

The term "the perfect storm" has become cliché in describing the current situation facing the forest industry in Canada (e.g., Anderson and Luckert 2007). Several factors have occurred simultaneously that are causing severe corporate losses and have led to numerous plant closures and thousands of unemployed workers across Canada (Natural Resources Canada 2007). If the factors lining up against the forest industry were short term, driven by the business cycle, these conditions would not present a convincing case for changing fundamental aspects of operating environments such as forest tenures. But several features of today's forestry landscape suggest that it will not be business as usual after the current economic downturn is reversed.

A key feature of current ills in the forest products industry is the declining US dollar. Between 2002 and January 2007, the US dollar value of spruce-pine-fir composite price index was almost constant, with prices in these two periods of $261/mbf and $269/mbf, respectively (*Random Lengths* 2008). But the same index expressed in Canadian dollars dropped from $418/mbf to $316/mbf – an overall decrease in lumber values of 39.3 percent. Although exchange rates may be cyclical, mounting US debt suggests that the strength of the Canadian dollar may not be a temporary state.

Another important feature of the current business environment is the fundamental shift happening in timber production. The global forest products industry is evolving rapidly, from a historical dependence on harvesting old-growth forests to wood supplies derived from intensively managed plantations. Canada's historic comparative advantage came from relatively cheap access to plentiful existing fibre sources. Although much of the forested area in Canada does not contain high-value trees, especially after centuries of harvesting, the pure size of the available fibre source has been a boon to the development of Canada's forest industry.[2] But given Canada's high northern latitude and corresponding slow growth rates of trees, it is not clear how Canada will fare in a global trade environment where profitability not only depends on cheap harvesting costs but also on cheap growing costs.

Moreover, the fibre that the forest industry has historically used to produce forest products may be harder to secure in the future. For example, Roberts (2008) refers to the emerging situation where the production of food, fibre, and fuel competes for land. Emerging scarcities and evolving technologies have created future scenarios that involve substitution possibilities between these land-based resources.

Competition for forestland may also come from other sources. For example, as discussed in Chapter 1 and in later parts of this chapter, the security of access to wood fibre is being brought into question by ongoing Aboriginal land claims. As negotiations and agreements evolve, the role of forests (i.e., who uses forest resources for what purposes) may shift markedly. Also, forests have seen increasing demands for non-timber forest resources, including non-consumptive uses such as recreation and consumptive uses that include a great variety of plant and animal products. There are also increasing public concerns for aspects of forests that are not directly consumed, such as their role in maintaining biodiversity and their ability to sequester carbon.

Although not all of these demands compete with fibre production, these values do, at a minimum, complicate fibre management decisions. Whereas historical management decisions were largely based on timber management objectives, forest planning is increasingly required to facilitate considerations regarding multiple resources valued by multiple stakeholders. As discussed in Chapter 1, these varied demands have led to fundamental changes in the governance of forests, including the rise of alternative forest policy frameworks – for example, certification. Consequently, forest companies are having to produce management plans to serve multiple, and sometimes conflicting, standards (Golec and Luckert 2008).

An additional challenge for the forest industry involves trade disputes with the United States over softwood lumber. With the United States being the primary market for Canadian lumber, these disputes and their resulting trade barriers remain an ongoing concern to the Canadian lumber industry. Over the past three decades, Canada has been involved in recurring court cases and negotiations with the Coalition for Fair Lumber Imports in the United States. A recent memorandum of understanding to arise from this process includes provisions that increase Canada's export tariff to the United States as lumber prices fall. Therefore, the effects of decreased housing starts, due to the recent recession in the United States, has been exacerbated for Canadian producers because of increased trade barriers during a time of bad markets (Zhang 2007).

The market situation for lumber is made even worse by current infestations of mountain pine beetle that are killing large areas of lodgepole pine in British Columbia and are spreading eastward into Jack pine stands across the country. The increase in harvestable volume arising from "beetle-kill wood" has led to what is now often referred to as a "wall of wood." This influx of wood, largely trapped in Canada behind the softwood trade barrier, has few places to go and could therefore be a contributing factor to low softwood lumber prices.

The above factors taken together suggest that the Canadian forest products industry is facing challenging current and future times. The relevant questions for the purposes of this book are, what has the tenure system done to contribute to these problems, and what changes to forest tenure policies might help alleviate them?

Current Problems with Forest Tenure Systems and Potential Future Directions

In the previous chapters, we discussed how individual characteristics and attributes of forest tenure systems, forest planning and practices, and stumpage fee systems affect the ability of forest policies to move forest management toward the objectives of sustainability. These individual characteristics make up forest tenures that have led to some general problems with current systems. The general themes behind these problems include undue focus on sustained yield of timber, low comprehensiveness of rights to forest resources, forced vertical integration, problems associated with land use decisions, and problems surrounding command and control approaches to forest governance. It is to each of these interrelated factors that we now turn.

Undue Focus on Sustained Yield of Timber

The forest management paradigm of sustained yield has been remarkably successful. It has been the framework within which forest management has been practised for over a century on many parts of the globe. But the focus of sustained yield on maintaining timber volumes has come under increasing attack as concepts of sustainable forest management have evolved.

Luckert and Williamson (2005) have reviewed a number of criticisms of sustained yield that have arisen in the forest economics literature. To begin with, sustained yield has been shown to be costly. Cut controls that cause firms to overproduce during low markets and to underproduce during strong markets can severely constrain potential benefits to producers and consumers (e.g., Boyd and Hyde 1989). But much of the initial logic behind sustained yield was not based on industry profitability. Rather, the concept was predicated on objectives of community stability (e.g., Le Master and Beuter 1989). However, studies have also revealed that sustained yield can actually be destabilizing. Cut controls that force firms to overproduce during poor markets and underproduce during strong markets can amplify price fluctuations – exacerbating, rather than stabilizing, impacts of business cycles (e.g., Dowdle 1984). Sustained yield harvesting constraints can also influence investments in new crops. When annual allowable cuts are calculated, investments that increase future yields may permit forest companies to increase current harvests of mature trees through the allowable cut effect.[3] As a result of the allowable cut effect, rights to harvest current volumes, rather than future benefits created from planted trees, tend to direct investments in regeneration, creating a wedge between forest management activities and the creation of a desired future forest (e.g., Luckert 2001). Finally, the timber focus of sustained yield creates situations where provisions for non-timber products are treated as constraints, rather than as objectives in and of themselves (e.g., Bouthillier et al. 1992).

The basic underlying tenet of sustained yield is well illustrated by the words of Abraham Knechtel, an early Canadian forester who said that the motto of the Dominion Forest Reserve was: "Seek ye first the production of wood and its right use and all these other things will be added unto it" (Murphy et al. 2002, 11). In essence, the logic behind sustained yield was that timber fibre was a sufficient indicator, as part of a coarse-filter approach, for managing multiple forest resources. But movements toward sustainable forest management have shown us that this course-filter approach is no longer sufficient. Instead, sustainable forest management is about paying specific attention to non-timber forest resources, rather than assuming that they will follow timber production in sufficient amounts. Focusing on multiple resources implies identifying complementarities and trade-offs among values that forests provide. Moreover, the complexities inherent in such analyses have pointed toward the need to consider adaptive forest management that involves structuring management so we can learn and adapt as we proceed.

Although most practitioners in forestry would claim that we are now past sustained yield and well into the world of sustainable forest management, policies that entrench sustained yield in forest management are still prevalent across Canada. Harvest volumes in every province are still regulated by allowable annual cuts that are central to forest management planning. In this country's largest timber-producing province, British Columbia, constraints designed to minimize the impact of environmental regulations on allowable annual cuts are enshrined in law (see Chapter 4). As such, a key question is how to change the emphasis on sustained yield as a component of forest tenure arrangements and replace it with new requirements that facilitate sustainable forest management.

Luckert and Williamson (2005) argue that sustainability constraints are needed in cases where markets for forest resources are missing, and/ or in cases where it is difficult to assign property rights to certain types of forest resources. They go on to say that, "with SY policies, we have probably chosen to attach strong sustainability policies to the only forest resource that does not need such protection (i.e., timber), while we have excluded other resources that could well need such protection (e.g., biodiversity) for pursuing SFM" (356).

In short, if we wish forest tenures to facilitate a broader range of goals, we will have to change their structures to match the new objectives being sought. This will involve refocusing policies on management for multiple outcomes by restructuring forest tenure agreements in terms of the rights they grant and the requirements they exact. In the next section, we consider rights and, subsequently, requirements.

Low Comprehensiveness of Rights to Forest Resources
As discussed in Chapter 3, Canadian forest tenures generally grant fairly narrow rights: rights to harvest trees, frequently individual species of

trees. These rights are in stark contrast to the breadth of resources that sustainable forest management implies (Luckert and Boxall 2009). We find, then, that industrial firms holding forest tenures are expected to manage for a breadth of activities that far exceeds the narrow rights that tenures grant. One means, therefore, of aligning current tenures with sustainable forest management objectives is to increase the comprehensiveness of forest tenures. We begin this discussion by concentrating largely on increasing rights to forest resources that are traded in markets: rights to grow trees, rights to multiple timber species, and rights to oil and gas. We then conclude this section by considering rights to non-marketed values.

Rights to Grow Trees

In Chapter 2, we discussed how tenures are structured so that benefit streams derived from planting future forests are not clearly held by forest tenure holders. As a result of this situation, links between current management practices and the designing of desired future forests are tenuous. How then could such links be strengthened?

Before considering how a forest tenure may be structured to create rights to grow trees, it is useful to think of conditions that would provide such rights on private land. A clear right to grow trees would require the property holder to perceive a benefit stream from harvesting future forests that would exceed anticipated costs. In a private property situation, attracting capital would likely require discounted future benefits to exceed the costs of establishing and tending a new forest.

Contrary to the private property case, current forest tenures curtail future benefits and increase future costs in a number of ways. To begin with, future benefits are uncertain because durations of tenures are generally far less than the time required for trees to be ready for harvest. Moreover, the prospect of governments making changes to forest tenures during their terms, and upon renewal, can instil considerable insecurity in perceived future benefits. The situation is further exacerbated by the fact that in all provinces, other than British Columbia, tenures are not freely transferable and are frequently not divisible such that parts of tenures may be sold. As discussed in Chapters 2 and 3, current forest tenures also imply significant costs associated with having to abide by requirements that include planning and restricted forest practices. Also, as discussed in Chapter 5, current stumpage systems are not structured to consider costs of establishing new forests, and may therefore preclude incentives to create such forests (Luckert and Haley 1990). Given these barriers to creating growing rights, it follows that rights could be created by lengthening the period of tenures, improving transferability and divisibility, assuring tenure holders that tenures will not change in ways that curtail their rights to future crops, reducing costs of current constraints, and charging stumpage based on the productivity of the land rather than on the value of current existing stocks. In other words, to establish growing rights, it would likely be necessary to come close to simulating private property on public land.

But even if forest tenures were structured to facilitate clear rights to future benefits, the underlying finances of investing in trees could preclude the creation of rights. Unfortunately, the financial returns to growing trees over much of Canada, particularly the vast boreal forest, is negative (e.g., Benson 1988; Rodriguez et al. 1998).[4] Therefore, a strict financial appraisal of reforestation would suggest that active reforestation is not worth the cost in many areas. But public sentiment would not likely tolerate harvesting trees without replacing them. What, then, is the financial calculation missing? Although it is beyond the scope of this chapter to consider all potential omissions and potential ways to internalize these omissions, some possibilities arise from previously introduced considerations.

In a sustainable forest management context, perhaps the key failing of private investment decisions is their failure to account for the breadth of values associated with forests beyond timber. Explicit consideration of values such as the maintenance of biodiversity or sequestering carbon would likely be absent from considerations of private firms. If private firms do fail to recognize all of the social benefits associated with reforestation, policy tools could be structured to incorporate these benefits into forest tenures. For example, Haley and Luckert (1998) suggest share cropping as a means whereby private firms and governments split input costs and output benefits, where private shares are made up of wood volumes and public shares are made up of public goods such as carbon values and biodiversity. If carbon markets were to develop, public shares that include the benefits of carbon sequestration would not likely be necessary, as private firms could benefit from marketed carbon.

Rights to Multiple Timber Species

As discussed in Chapter 2, rights to harvest timber are frequently restricted to particular species. In many cases, rights to species growing intermixed in one stand are granted to multiple tenure holders, causing inefficiencies because of a failure to manage forests for their jointly produced resources. If comprehensiveness were greater, a single firm could be responsible for the interrelated management decisions associated with growing multiple species on one piece of land.

Perhaps the largest barriers to increasing the comprehensiveness of timber harvesting rights are historic allocations that matched specific species to specific mills. Although problems associated with this practice will be discussed at more length below under considerations of vertical integration, it is important to note here that tenure reforms could allow one firm to harvest and manage all species, and then sell logs to whatever processing plant would value them most. Such an approach, though, would require a means to consolidate harvesting rights to multiple species into the hands of single firms. If allowed, this consolidation could be accomplished with sales of rights among firms, where sales would be triggered by desires to avoid current inefficiencies associated with inter-firm coordination difficulties.

Rights to Oil and Gas

Inefficiencies associated with non-coordinated management have also arisen between forest companies and oil and gas companies. Such problems have historically been concentrated in Alberta, but new energy discoveries in British Columbia and Saskatchewan could cause this to become a more widespread problem.

Although forestry and oil and gas operations are not generally combined within single firms, there are some large energy firms involved in the management of forests (e.g., Shell Oil). Increasing comprehensiveness of rights such that oil and gas and timber resources were packaged together could potentially aid in the joint management of these resources. Under a single firm, road building, harvesting, and regeneration decisions could be coordinated with fossil fuel withdrawals.

But such an approach would likely face obstacles. Companies that produce oil and gas are typically specialized in such activities, working on much different time scales than companies involved in forestry. Moreover, the relative magnitudes of the dollars associated with oil and gas versus timber could result in a disregard for forest management. Policies that protect non-marketed forest resources would have to be carefully designed to weigh the trade-offs associated with such large commercial potential. But under current regulatory frameworks, policies to protect forests from oil and gas exploration are not likely to be forthcoming. Current oil and gas exploration activities face few controls regarding disturbances to forests because their activities do not fall under the jurisdiction of provincial forest acts, which have as one of their expressed purposes the protection and sustainability of forest resources. If oil companies were to acquire forest tenures, all of their forest-based operations could potentially fall under the scrutiny of the various forest acts under which forest tenures are granted – a complication that private firms would likely try to avoid. But given that forest management objectives can seriously conflict with fossil fuel operations, policies that cause these trade-offs to be ignored are avoiding potentially serious policy issues.

Rights to Non-Marketed Forest Resources

Thus far, discussions about expanding rights to forest resources have largely been confined to resources sold in markets. But, as discussed in Chapter 2, many forest products are not sold, largely because it can be difficult to establish exclusive property rights to some types of benefit streams. Some of these non-marketed resources could potentially be marketed. For example, some types of forest-based recreation (such as hunting), though not privately sold in provincial forests, are sold in other jurisdictions – for example, on private land in the United States and on public and private forestland in many European countries. But, as discussed in Chapter 3, private market transactions could favour commercial species at the cost of non-commercial species, thereby failing to facilitate biodiversity objectives. For other types of non-marketed forest resources, it could be possible for governments to establish markets as a type of regulatory tool. For example, Weber and Adamowicz (2002) discuss the potential to establish tradable permit markets that would limit areas of disturbances on forestlands.

The general problem to consider is whether markets, either privately formed or created by governments, cause private firms to make decisions that facilitate the various and changing objectives of sustainable forest management. To the extent that markets can allow private firms to make such decisions, such an approach is frequently desirable, as it harnesses the resources of the private sector to pursue social objectives in a flexible manner. To the extent that market forces cannot be harnessed in the public interest, we are left with situations where governments will likely have to retain rights to such resources and manage them on the public's behalf.

Forced Vertical Integration

Several characteristics of forest tenures, discussed in Chapters 2 and 5, have led to a vertically integrated forest industry. First, appurtenancy requirements require holders of forest tenures to operate a forest-product processing facility as a condition for receiving tenure. Such requirements preclude the possibility of a firm being solely in the tree growing business. Second, log export restrictions constrain transfers of logs between companies across provincial and international borders. With log markets constrained, firms may have few choices other than to process the logs themselves. Lastly, stumpage systems frequently value logs based on the end product for which they are destined. If the value of a log depends on what it will be used for, firms will not have incentives to find the most profitable use of a log. Instead, such stumpage systems cause logs to be destined to specific manufacturing processes.

The mixed-wood management problems discussed above are symptomatic of the issues associated with forced vertical integration. In short, policies that tie logs to specific manufacturing processes diminish the capacity of the industry to adjust to changing technologies that alter the potential values of logs. In the case of mixed-wood management, such rigidity causes governments to address technological change by allocating new, overlapping tenures that tie newly valued species to yet another production process.

The problems associated with mixed-wood management could well represent the tip of the iceberg compared to future issues discussed above regarding food, fibre, and fuel. In the context of such change, policies that connect specific resources to specific processing technologies will face restricted opportunities and thereby reduce the value that these raw materials may contribute to economies. In order to have a forest-based economy that is more flexible in response to changing conditions, it follows that links between forests and specific end products should be eliminated. Therefore, options for eliminating appurtenancy requirements, loosening log export restrictions, and eliminating stumpage fees based on end-product prices could be considered. Although wood-product plants would then have to compete with other processors, such competition is vital to assure a dynamic and vibrant industry, and to allow society's forest resources to find their highest values.

Problems Associated with Land Use Decisions

The need to consider trade-offs and complementarities on forestland bases has become increasingly evident in discussions of land use decisions. In particular, alternative zoning strategies have been part of discussions about the pursuit of sustainable forest management. By better matching land uses with goods and services that society desires, zoning provides the potential for us to get more from our forests (e.g., Vincent and Binkley 1993; Sahajananthan, Haley, and Nelson 1998). One such zoning strategy is referred to as "Triad," whereby forestlands are divided into protected, extensive, and intensive management zones. But operationalizing Triad has been elusive. Specifically, there has been little progress on how these zones would be determined within the context of current tenure systems.

Historically, most zoned areas in provinces have been specified by governments at a very large scale. For example, in Alberta, there are green and white zones that distinguish respectively between publicly owned forests and privately owned agricultural operations. However, strategies such as Triad require a much higher degree of resolution in weighing the costs and benefits of land use decisions, and will also likely require flexibility to change these decisions over time. Therefore, the Triad strategy could potentially benefit from allowing private firms to make land use decisions within larger zones established by governments.

Anderson (2008) has recently investigated such a situation where tenure holders are allowed to make land use decisions within the context of a combined public and private land base. A model is constructed where tenure holders make decisions on whether to continue extensive practices as usual, intensify management with fast-growing hybrid species, or retreat from areas that are less profitable, thereby establishing de facto protected areas. The analysis shows that allowing firms the flexibility to make land use decisions can result in large increases in wealth, derived partially from withdrawing from low-return areas, thereby increasing the protected zone. Some of the important policy considerations in structuring such a system include the role of cut controls and whether private land investments can be used to invoke the allowable cut effect on public lands, whether withdrawing from low-return lands creates valuable protected areas or instead is interpreted by governments as underutilization of harvesting rights, and on what types of land fast-growing species are allowed.

Problems Associated with Command and Control Approaches

As indicated in Chapter 4, a dominant feature of Canadian forest tenures is the various requirements and control systems that have emerged in an attempt to protect non-timber resources by constraining forest harvesting rights. A number of problems have arisen as a result of such policies.

First, as mentioned above, the costs of these constraints can be a major factor influencing the profitability of forest companies (G.C. van Kooten 1994; Haley 1996; Clarke 1997). For example, Pearse (2001) shows how the introduction of the Forest Practices Code, a policy in British Columbia that added numerous restrictions to timber harvesting, was associated with a decline in the invested capital stock in coastal operations.

Second, as discussed in Chapter 4, it can be difficult to develop requirements that are congruent with local ecological and social situations. Heterogeneity in social and natural features of forests can make it difficult to establish requirements that are relevant. In response to such concerns, requirements are frequently established that vary depending on local situations. But more variability in the specification of requirements can increase the complexity of administration and may still not provide sufficient flexibility to meet the many different conditions of forests.

Third, such constraints can lead to undesirable incentives for forestry firms. In the context of requirements, firms have incentives to minimize the costs for meeting a requirement, while considering strategies of avoiding penalties that may be imposed if they are caught not complying. A major problem of such incentives arises because benefits that result from required activities are external to the forestry firm (Luckert 1998). For example, the absence of growing rights, discussed above, occurs because forestry firms do not have clear rights to benefits from growing trees. In the absence of growing rights, provinces have put in place regeneration requirements and systems designed to enforce such provisions. With benefits of requirements absent from firm decisions, potential large welfare losses arise from firms that "just do what they are required" rather than having the freedom to consider alternative means of achieving alternative levels of benefits. Efficiency implications of such restrictions are further exacerbated by disincentives to invest in innovation. For example, with benefits of growing trees external to forestry firms, there is little incentive to invest in research to improve characteristics of trees that will be harvested in the future.

One approach to addressing these types of inefficiencies has been to consider variants in the specificity of requirements (see Chapter 4). For example, objective-based approaches, attempted in some jurisdictions, including Alberta and British Columbia, have sought to decrease the specificity of requirements, thereby increasing the discretion of firms. But such approaches have faced problems. Even if requirements allow firms freedom to pursue higher-order objectives and to propose alternative means of meeting these requirements, governments must still decide how to monitor and enforce progress toward the more general objectives. This process of establishing intermediate monitoring steps can sometimes inadvertently cause the system to revert to being highly prescriptive, especially if governments do not believe that firms have incentives to faithfully pursue the objectives. Key questions about the structuring of such approaches include identifying under what conditions increased flexibility may be able to be maintained in order to provide tenure holders more choice when adopting management strategies.

Principles for Change: Essential Attributes of Forest Tenures Designed to Further Sustainable Forest Management

The above summary of current problems associated with forest tenures and potential solutions discloses several common threads, which, taken together, may be interpreted as principles for change. In other words, if we were to set out to design a forest tenure system that attempted to achieve the objectives of sustainable forest management while recognizing the rapidly changing domestic and global socio-economic environments in which Canada's forest resources are managed, what attributes would this system have to embody?

First, the system will have to provide for the *integrative management* of jointly produced resources. Pressures from multiple uses and users of forestland have increased, and such trends are likely to continue. The independent use and management of these multiple interrelated resources can lead to inefficiency and confrontation. Future tenure frameworks must provide an environment where trade-offs and complementarities between forest-based resources can be considered and managed for. In some cases, this may call for a tenure framework that facilitates harvesting and management decisions by private firms based on market values. In other cases where markets are likely to fail, tenures may have to be changed so that governments take a more active role in facilitating the provision of ecological and cultural values.

Second, the forest tenure system will need to provide a framework that allows *flexibility* to respond to variable and changing circumstances. The heterogeneity regarding social and natural features embodied in various sustainable forest management approaches, and the dynamic nature of forests and values, require flexibility to change management decisions within a given forest tenure framework, and to change the framework over time. Adaptive behaviour is a key element of sustainable forest management and must be facilitated and encouraged, rather than constrained by rigid policies. But flexibility does not imply giving governments carte blanche to continuously change policies, thereby eroding commercial investment incentives. Although changes to government policy are inevitable, if policy frameworks allow companies room to adapt to changes, then policy changes may be farther and fewer between.

Third, the forest tenure system should provide incentives for *innovation*. Innovation allows us to be smarter in pursuit of economic, ecological, and social sustainable forest management objectives. In order to adapt to changing conditions, innovation will be required from industry, government, and other interested publics. If forest tenures do not facilitate such innovation to occur, it will become increasingly difficult to pursue and deliver on social objectives.

Finally, forest tenures should seek to provide *clarity* regarding rights and responsibilities of all stakeholders – private and public. Clearly defined rights and responsibilities regarding the management and planning of current and future forests for multiple forest resources will be necessary not only to create incentives for effective management but also to establish accountability. As forest resource management issues have become increasingly complex with the broad scope of sustainable forest management objectives, clarity has become increasingly important.

Barriers to Change

There are many indicators that the Crown forest tenure system in Canada is not adequately achieving the objectives of sustainable forest management. Indeed, a conversation with almost any forest industry executive or senior provincial bureaucrat will affirm this impression. Many of the problems and their potential solutions, some of which are the subject of this book, are fairly well understood. Yet, with the exception of British Columbia, provincial governments have been reluctant to consider fundamental, systemic changes to their forest tenure systems to reflect the reality of rapidly changing domestic and global social and economic environments. Why is this? In this section, we briefly explore perceived barriers to change, and in particular four types of barriers: intellectual barriers resulting from the challenges of articulating and mobilizing political support for a clearly superior alternative to the status quo; political barriers created by those who perceive relative benefits in the status quo; decision rules that empower opponents to change; and a mismatch between reform-minded jurisdictions and the broader institutional environment of tenure policies.

Intellectual Barriers

There is widespread agreement that the existing tenure system is flawed, but there is no comparable degree of agreement, even among policy specialists, about what would constitute superior alternatives to the status quo. Some analysts are inclined to emphasize increasing private rights, whereas others see the solution in decentralization or other alternatives. This lack of agreement stems from two sources. First, different analysts may hold different values about what the most important consequences of policies are; some place priority on improving economic efficiency, others might place greater priority on environmental effectiveness or concerns about social justice.

Second, there are significant uncertainties about the outcomes of alternative approaches to reforming tenure. So even if analysts were able to agree on how to evaluate what would constitute an optimal tenure, they might disagree about what the consequences of different alternatives would be. Although there is a considerable body of theory concerning the impact of tenure reforms, empirical evidence, in most cases, is sparse, creating a great deal of uncertainty around possible outcomes. Forest tenures are components of complex economic and social systems within which patterns of cause and effect are difficult to identify, and human behaviour, in the face of change, is notoriously difficult to predict. In forestry, these challenges can be particularly formidable given the significant temporal delays between policy or management, interventions, and evidence of system response.

For example, it is theorized that extending forest tenure rights to marketable non-timber products may result in the production of multiple forest products in a coordinated, efficient manner – reducing regulatory compliance costs and facilitating sustainable forest management. However, as we have seen above, the extent to which such a policy is successful will depend on many factors, including the opportunities for managing for non-timber values within the forest ecosystems covered by the tenure arrangement; the nature of trade-offs among the products concerned – particularly timber and non-timber products; the costs of generating inventory data and the accessibility of technical know-how; the availability of technical expertise; and the presence of viable markets for the products concerned and their relative prices. Even if such a policy change is successful, there may be unforeseen circumstances that may be costly both ecologically and politically. For example, a licensee provided with the right to manage for wildlife may decide to do so. However, if revenues mainly flow from the sale of hunting rights, incentives may be created to eliminate species such as cougars or wolves that prey on large ungulates, with serious negative impacts on biodiversity. Or plans to increase deer populations may see a trend toward larger clear-cuts and burning. Under these circumstances, vigorous opposition could be expected from environmental NGOs.

Another example of the challenges of uncertain outcomes comes from the issue of privatization. Much of the rationale for increasing privatization of rights to forestland is based on the belief that it would increase yields of forest products, especially timber but perhaps others as well. With greater security of rights, managers of forestland would be able to benefit from increased investment in enhanced silviculture. In reality, there are uncertainties about what the financial returns would be to a variety of silvicultural investments. There are also significant uncertainties about whether tenure holders would actually choose to make these investments in the growing conditions they face across Canada. If policy makers choose to privatize some public land on what turns out to be an erroneous assumption about the magnitude of benefits, the costs of the policy change might not have been justified.

Political Opposition from Those Benefiting from the Status Quo
The second category of barriers to change arises from potential political opposition. Public policies can be considered bundles of costs and benefits to different interests in society (J.Q. Wilson 1980), so policy change inevitably changes that distribution of costs and benefits. In other words, any meaningful policy change creates losers as well as winners. As we have pointed out throughout this book, tenure systems have

multiple goals. Each set of goals may serve a somewhat different group of stakeholders, although there is some overlap. Thus, environmental goals serve the interests of environmental NGOs plus an increasing proportion of the general population. Distributional goals are largely the focus of labour, forest-dependent communities, and increasingly, First Nations. Economic goals, although certainly important to labour and local communities, are of major interest to and promoted by the forest industrial sector. These goals frequently conflict, and policy changes designed to serve one goal – improve forest industry efficiency, for example – may seriously compromise the system's ability to meet other goals.

Stakeholders who anticipate relative losses from policy changes will naturally oppose those changes. Relevant actors include not just interest groups outside government but also bureaucratic officials and organizations whose status or influence may be negatively affected by policy change. Bureaucratic resistance is likely to be especially acute when significant reorganization of government functions is required to implement policy change. In provinces like British Columbia, where forest and related resource management is distributed across multiple agencies, changes to effect integrated resource management face formidable challenges.

Politicians are reluctant to pursue policies that impose costs on influential groups, even in circumstances where the changes may bring about net benefits to society. These political barriers to change are compounded by the intellectual barriers described above. Although the benefits of proposed policy change are frequently speculative, interest groups are especially skilled at making their version of the costs of change highly certain to politicians. That the benefits of policy change may appear only in the future, far beyond a politician's electoral time horizon, whereas the costs of change may appear in the short term, may further increase a politician's reluctance to depart from the status quo.

Decision Rules That Advantage Opponents to Change
In some important cases, decision rules can increase the barriers to policy change if they grant opponents to policy change the ability to block change, or at least make it more difficult. For example, as we describe in

more detail below, some provinces require governments to compensate tenure holders for a reduction in their timber allocations beyond some specified minimum amount (in British Columbia's Forest Act, 5 percent of the allowable annual cut). In addition, Aboriginal groups have, through a series of actions by Canada's courts, been granted augmented rights that increase their influence over changes that affect Aboriginal rights and title, including treaty rights. As we described in Chapter 1, the most prominent example is the Supreme Court of Canada's 2004 decision in *Haida*, which ruled that the government did not have the authority to transfer a tenure from one company to another without adequately consulting and accommodating the interests of the Haida Nation. Although the companion *Taku* case clarified that this obligation for government does not constitute a veto for Aboriginal groups, it certainly empowers Aboriginal groups to resist changes to tenure.

Institutional Mismatch

In certain circumstances, policy change might also be constrained by a mismatch between the jurisdiction of the government intent on change and other governments. Constitutionally, the federal government of Canada has jurisdiction over marine fish and wildlife and fish in streams and lakes that migrate between salt and fresh water. Thus, in designing forest practices regulations that affect the habitat of migratory fish species, provinces are constrained by federal regulations. In fact, during the 1970s and 1980s, the federal Fisheries Act and regulations provided powerful stimuli for the introduction of provincial measures to protect freshwater environments from forestry and other land uses. The federal government also has jurisdiction over migratory birds, and other responsibilities resulting from its international treaty powers. If the federal government is at odds with a reform-minded province, these jurisdictional levers could be used to thwart change.

The ongoing dispute with the United States over Canadian exports of softwood lumber could also restrict government policy responses. If governments must continuously consider proposed changes in light of their implications within the context of countervailing duties, there will likely be some options that are not feasible given the political economy of trade negotiations. Indeed, the 2006 Softwood Lumber Agreement has an anti-circumvention clause that prevents Canadian governments from taking policy measures to offset the effects of the export restrictions contained in the agreement. The effect of this provision is to impose an American softwood lumber lobby screening on all significant forest policy reforms (Zhang 2007). However, there may also be a positive aspect to current trade environments in facilitating change. Governments, which might otherwise face obstacles from inertia discussed above, may need outside motivation to consider policy changes. For example, many of the recent policy changes in British Columbia may not have happened but for the pressures associated with trade negotiations.

Path Dependence

As a result of these intellectual, political, and institutional barriers to change, major policy changes such as radical reforms to provincial forest tenure systems are rare. Consequently, institutional frameworks tend to lag behind changing economic circumstances and fundamental shifts in social norms and aspirations. Scholars refer to this phenomenon as "path dependence," a situation in which present options are severely constrained by past decisions, even though the circumstances that gave rise to those prior decisions are no longer relevant (Pierson 2004). Once a policy or institutional path is established, entrenched mindsets, interests, and institutions make departures from the status quo difficult to envision. Extant forest tenure policies in Canada are a classic case of path dependence.

When major changes in institutional frameworks do occur, it is usually in response to an abrupt combination of changes, such as a new, ideologically motivated political party in power at the same time as a crisis situation when the political, social, and economic costs of inaction outweigh the benefits of maintaining the status quo. Such was the case in New Zealand in the mid-1980s when, in response to a severe fiscal crisis, a socialist government introduced a program of reforms designed to free the economy of public intervention, dismantle the welfare state, and place increasing reliance on free-market mechanisms as a means of allocating resources and distributing wealth. In the forest sector, this involved privatizing a public forest estate that comprised about 50 percent of the country's industrial forest plantations. Likewise, the major changes in forest policy that were introduced by the market-oriented BC govern-ment in 2003 as part of its Forestry Revitalization Plan were motivated by a perception that the province's forest industry, particularly on the coast, was facing problems of crisis proportions (Pearse 2001) that threatened the economic welfare of thousands of British Columbians.

Overcoming Barriers to Change

In a general sense, change should occur if we have better options than those we are currently following. But the barriers presented above could preclude us from identifying such situations and from instigating change. How, then, can these barriers be overcome? In the following sections, we consider the need for making the case for change; being careful with, and learning from, change; and cushioning the effects of change: the issue of compensation.

Making the Case for Change

Many of the barriers to change listed above could be overcome if stake-holders were convinced that better options exist than those we are cur-rently following. Toward that end, this book has sought to provide information about the current status of forest tenure policies and options for future policies. Although it is beyond our scope to delve into the particular policy objectives and performance of each province's tenure system, we have identified a number of specific features, common to many current tenure policies across Canada, for consideration in po-tential tenure reforms.

But even if ideas proffered by academics are accepted by politicians, little will happen unless politicians perceive that they have secured the social licence necessary to enact change. Social licence is frequently sought through complicated processes involving multiple stakeholders. It is during these processes that the case for change must be made with compelling reasoning and clarity.

Perhaps the most compelling and clear expression to guide change comes from an adage frequently used by architects: "form [ever] follows function" (Sullivan 1896). Following this maxim, the key question that must be asked is, what values are tenure policies trying to facilitate? As replies to this question are generated, three major trends and resulting implications for tenure policy emerge. First, we know that values derived from forests have changed significantly since tenures were first designed. Therefore, the time is ripe for us to consider changes to forest tenures. Second, we know that values also differ significantly between regions and forests. Therefore, we will likely need forest tenures that vary significantly to match the spatial heterogeneity of our forests. Lastly, we know that values are likely to change further over time. Therefore, as also discussed above, flexibility in designing future tenures will be key.

Being Careful with, and Learning from, Change

Despite the case for tenure reform, enthusiasm for change should be tempered with humility associated with limitations in our abilities to design effective policies. Despite our discussions of the implications of alternative tenure structures, specific natural and social local conditions make generalization difficult. Moreover, as mentioned above, there is potentially great uncertainty associated with changing forest tenures. We simply do not have enough experience that links specific policies to specific results to allow us to predict exactly what will happen when policies are changed in a given location. Policy changes are notoriously plagued with unintended consequences. Therefore, the case for being cautious with change is convincing.

One way to exercise caution regarding change is to experiment with small changes before changing policy en masse. For example, in some cases it may be prudent to introduce changes incrementally, perhaps one at a time, rather than changing numerous aspects of a tenure system simultaneously. Such an approach will provide analysts with a better chance of isolating impacts of policy changes. But interpretations of such experiments should be made cautiously, as incremental changes in policy may lead to differing impacts depending on the specific policy framework in which the change is made. Another way of introducing marginal changes is to experiment with larger changes in tenure policies but on a smaller scale. For example, experiments with increasing the comprehensiveness of forest tenures could be piloted within specific areas and by adding specific rights.

However, in some cases, small changes in policy may not produce desired results. For example, above we point out that appurtenancy requirements, log export restrictions, and stumpage fee systems based on end-product prices are likely all contributing to vertical integration. Therefore, changing just one item from this list may have minimal effects. Similarly, current economies of scales in forest harvesting, management, and processing may preclude the financial viability of small-scale experiments. And finally, if we wish to speed up the learning that may accompany policy changes, we may wish to design experiments in policy changes that look at more drastic changes.

Designing experiments to learn from change is central to concepts of adaptive management that have become an integral part of pursuing sustainable forest management (e.g., Burton et al. 2003). However, concepts of adaptive management have most often been used in the context of the natural sciences, where, for example, impacts of alternative forest management activities have been explored. The case for extending these concepts to social science experiments is strong. Despite the many forest policy options that provinces across Canada have tried, there is a dearth of information regarding what policy experiments have been conducted. Moreover, these efforts are rarely monitored in ways that allow for the assessment of strengths and weaknesses of policy changes.[5] In short, there is a need to design and monitor policy experiments much in the same way as is done in natural science research.

Cushioning the Effects of Change: The Issue of Compensation

Complications that governments face in designing forest tenure systems point toward the need for constant review and change. In other words, complications call for the adaptive management of forests. But changes in forest policy almost inevitably benefit some stakeholders while imposing costs on others. When some stakeholders are made worse off by changing forest policies, significant barriers to policy change may arise. But part of policy reform could include forms of compensation for those harmed by changes. We therefore consider the question, should the Crown pay compensation for changes in forest policies?

In addition to being aware of needs for policy changes, governments are also generally well aware of the benefits of stability. Stability in forest policies may enhance the incentives for private firms to invest in renewing forests and maintaining processing facilities (Luckert 1991b).

For example, Pearse (2001) discusses how capital has left the BC coastal forest industry in response to an uncertain policy environment. We see, then, that policy makers are faced with having to make trade-offs regarding the need to adjust policies for changing social conditions (i.e., social goals and market conditions), while recognizing that creating insecurity for tenure holders could add uncertainty to business environments.

One means to allow governments flexibility to change policy yet maintain secure business environments is for the government to offer, or for the courts to award, compensation for changes in policy (e.g., Innes 1995). However, there is very little precedent for such action in Canada. In considering compensation for restrictions on private land-owners in the case of the Species at Risk Act, Pearse (2000) states: "But the courts and governments have historically drawn a distinction between expropriation of property, for which compensation is due, and restrictions on the use of property for some public purpose, for which compensation is generally not payable." In the case of forest tenures, the case for compensation to tenure holders seems even weaker, as the interests held by private firms are weaker than in the case of private property.

There is precedence in many provinces for changing tenure policy without compensation to tenure holders. This prerogative of the Crown reflects the view that tenures, or parts thereof, are not necessarily considered contracts. Instead, they are agreements subject to government policies that may change in ways that influence the benefits tenure holders derive. Nonetheless, there are some forest tenures for which clauses in legislation stipulate conditions under which compensation will be paid by the Crown. However, such clauses are uncommon and their scope is very limited, generally providing compensation only for significant reductions in allowable harvest levels and/or improvements, such as roads, to public forestland.

It is with good reason that governments have limited legislated compensation provisions for changing forest policies. Forest companies make investments in forests and processing plants in the face of numerous sources of uncertainty. These sources of potential change include changes in market conditions and changes in social conditions that may stimulate changes in forest policy. Logic that keeps governments from compensating private firms for losses due to market uncertainty may similarly be applied to considering policy uncertainty. In both of these cases, it may be advantageous for firms to have incentives to consider these uncertainties in their investment decisions, whether they be from markets or other changing social trends. To the extent that compensation clauses protect private firms from adverse changes due to changing social conditions, incentives to integrate these uncertainties into planning are eroded. Given that these uncertainties arise from changing social values, it is important that investing firms consider these potential changes in their decisions. Otherwise, protection from such uncertainty through compensation would lead to too much investment from a social point of view (e.g., Blume, Rubinfield, and Shapiro 1984).[6]

This logic regarding compensation follows the concept of mutability discussed in Chapter 3. With a vast majority of forestland under public ownership in Canada, provincial governments play a significant role in managing forest resources. The policies that enable this management have numerous objectives. The ultimate efficacy of these policies lies in their ability to further social objectives. Challenges to designing a tenure system include adjusting the system to changing social objectives and market conditions. Tenures change and tenure holders expect such changes and plan accordingly. Problems arise, however, when these changes create tenure insecurity. Such an investment environment may be created if changes to policy appear to shift haphazardly, based on shifts in political ideology. In such cases, firms are not provided with a stable framework within which social considerations may be adequately considered. Therefore, a key point in considering whether compensation should be paid is whether the Crown appears to have good reasons in its decisions on why and how changes to forest policies are made. Whether policy changes are "for good reason" will be judged by industry, investors, and the public at large. If policies arise from processes that are transparent and representative of stakeholders' interests, they are much more likely to be judged as being for "good reason." Open discussion among stakeholders is key. Conversely, if policies emerge unexpectedly from closed-door processes, the logic behind the policies may elude stakeholders and contribute to tenure insecurity.

In summary, forest policy across Canada has generally limited compensation to specific and large changes to tenure agreements (i.e., generally, for large reductions in the allowable annual cut granted by tenures). Although it is widely recognized that such changes can create investment uncertainty for tenure holders, they may nonetheless be necessary for the Crown to adjust to changing social objectives. Moreover, the payment of compensation for such changes may be undesirable in that it has the potential to lead to misallocated resources in the forestry sector. If forestry firms are buffered from political uncertainty stemming from changing social values, they will not have incentives to consider these uncertainties in their investment calculations and may over-invest. Such logic depends on the Crown being motivated by changing social objectives and thereby having good reason to change tenure policies.

Toward "Smarter" Policies for Sustainable Forest Management

This book reflects a history wherein numerous and ongoing changes have added layer upon layer of regulation to forest tenures as forest issues have emerged. However, a key theme is that, over time, there is a growing dissonance between consequences of the tangle of inherited policies and the values of contemporary society. The stakes have become too high, and the goals too complicated, for such an approach to work. Instead, we believe it is time for careful consideration to be given to new approaches of achieving sustainable forest management.

Such approaches will need to consider carefully how systems may be structured that allow for integrative management, flexibility, innovation, and clarity of rights. Achieving these attributes in a tenure system will require that each and every condition of tenures be carefully considered with respect to design, purpose, and costs versus benefits of alternative approaches. In short, we need a "smarter" approach where a more refined set of property rights are combined to better align policies with values in a particular time and place.[7]

Appendix
References for Details on Canadian Crown Forest Tenure
Characteristics, Forest Practices Regulations, and Stumpage
Systems by Province

British Columbia

Province of British Columbia. *Forest Act* [R.S.B.C. 1996] c. 156.

–. *Annual Rent Regulation.* B.C. Reg. 122/2003.

–. *Advertising, Reports, Disposition and Extension Regulation.* B.C. Reg. 55/206.

–. *Timber Harvesting Contracting and Subcontracting Regulation.* B.C. Reg. 22/96 (inc. Reg. 524/2004).

–. *Community Forest Regulation.* B.C. Reg. 352/2005 (inc. Reg. 255/2007).

–. *Minimum Stumpage Regulation.* B.C. Reg. 354/87.

–. *Cut Control Regulation.* B.C. Reg. 578/2004 (inc. Reg. 384/2008).

–. *Forest and Range Practices Act* [S.B.C. 2002] c. 69.

–. *Forest Planning and Practices Regulation.* B.C. Reg. 14/2004 (inc. Reg. 106/2009).

–. *Forest Practices Code of British Columbia Act* [R.S.B.C. 1996] c. 159.

–. *Forestry Revitalization Act* [S.B.C. 2003] c. 17.

British Columbia Ministry of Forests and Range. *The State of British Columbia's Forests.* December 2006.

–. *Timber Tenures in British Columbia.* 2006.

–. Stocking Standards (spreadsheet). 2006.

British Columbia Ministry of Forests and Range. Revenue Branch. *Interior Appraisal Manual (effective July 1, 2007)*. 2007.

–. *Coast Appraisal Manual (effective June 1, 2007)*. 2007.

British Columbia Ministry of Agriculture and Lands. Integrated Land Management Bureau. *A New Direction for Strategic Land Use Planning*. December 2006.

–. *Report on the Status of Integrated Land Use Plans in British Columbia*. October 2006.

Haley, David. A Bold Move Forward: The New Market Based Stumpage System Promises to Get Stumpage Rates Right. *The Truck Logger*, Winter 2004: 29-35.

Martin, P.J. Design of Regeneration Standards to Sustain Boreal Mixed Woods in Western Canada. *International Forestry Review* 7(2) (2005): 135-46.

Alberta

Province of Alberta. *Forests Act* [R.S.A. 2000] c. F-22.

–. *Forest and Prairie Protection Act* [R.S.A. 2000] c. F-19.

–. *Forest Resources Improvement Regulation* AR 152/97 (consolidated to 170/202).

–. *Timber Management Regulation* AR 60/73 (consolidated to 266/2003).

Alberta Ministry of Sustainable Resource Development. Public Lands and Forests Division. *Forest Management Agreements*.

–. *Forest Management Fact Sheets*. April 2008.

–. *Timber Production Monitoring*. Directive no. 99-02. August 1999.

–. Series of information pamphlets dealing with various issues, including timber dues and Crown fees, compliance, and enforcement.

Alberta Ministry of Sustainable Resource Development. Public Lands and Forests Division. Forest Management Branch. *Alberta Forest Management Planning Standard*. April 2006.

Alberta Environmental Protection. *Alberta Timber Harvesting and Operating Groundrules*. Pub. no. Ref. 71. 1994.

Saskatchewan

Province of Saskatchewan. *Forest Resources Management Act* [S.S. 1996] c. F-19.1.

Saskatchewan Ministry of the Environment. FMA Standards and Guidelines. *Weyerhaeuser Prince Albert Forest Management Agreement Area Standards and Guidelines.* May 2004.

–. *Weyerhaeuser Pasquia-Porcupine Forest Management Agreement Area Standards and Guidelines.* May 2004.

–. *Mistik Management Forest Management Agreement Area Standards and Guidelines.* May 2004.

–. *L & M Forest Management Agreement Area Standards and Guidelines.* May 2004.

–. Fact Sheet 9. *Forest Protection Responsibilities. Saskatchewan Fire and Forest Insect and Disease Policy Development.*

Global Forest Watch (Saskatchewan chapter). *Deforestation: Lack of Regeneration in Saskatchewan Forests.* Technical Report no. 1. March 2000.

Manitoba

Province of Manitoba. *The Forest Act* [C.C.S.M.] c. F150.

–. *Forest Use and Management Regulation* 227/88R 60/2009.

Manitoba Ministry of Conservation. Forestry Branch. Forest Management. *Manitoba's Crown Timber Pricing.*

–. *Timber Sale Auctions.*

Manitoba Natural Resources. Forestry Branch. *Timber Harvesting Practices for Forestry Operations in Manitoba.* October 1996.

Ontario

Province of Ontario. *Crown Forest Sustainability Act* [S.O. 1994] c. 25.

–. *Forestry Act* [R.S.O. 1990] c. F26.

–. General, O. Reg. 167/95 (amended to O. Reg. 186/07).

Ontario Ministry of Natural Resources. *Ontario's Crown Timber Pricing System.* 2007.

–. *Forest Management Planning Manual.* 1996.

–. *Forest Management Planning Manual.* 2004, with April 2007 addendum.

–. *Forest Operations and Silviculture Manual.* 2000.

–. *Old Growth Policy for Ontario's Crown Forests.* 2003.

–. *Forest Management Guide for Natural Disturbance Pattern Emulation,* version 3.1. 2001.

Quebec

Province of Quebec. *Forest Act* [R.S.Q. 1996] c. 4.1.

−. *Regulation Respecting Contributions to the Forestry Fund.* R.Q. c. 4.1 r.0.02.

−. *Regulations Respecting Forest Management Plans and Reports.* R.Q. c. 4.1 r.1.02 and r.9.

−. *Regulations Respecting Forest Royalties.* R.Q. c. 4.1 r.12 and r.2.

−. *Forest Protection Regulations.* R.Q. c. 4.1 r.1.1 and r.11.

−. *Regulations Respecting Standards of Forest Management in the Domain of the State.* R.Q. c. 4.1 r.1.001-1.

Ministry of Natural Resources and Wildlife. *Forest Resources Protections and Development Objectives.* 2004.

−. *General Forest Management Plans 2007-2012 Instrumental Document.* 2005.

−. *Forest Management Plans Essential Elements for Sustainable Development.* 2006.

New Brunswick

Province of New Brunswick. *Crown Lands and Forests Act* [S.N.B. 1980] c. C-38.1.

−. *Timber Regulation 86-160.*

−. *Clean Water Act* [S.N.B. 1989] c. C-61.

−. *Wetland and Watercourse Alteration Regulation 2003-16.*

New Brunswick Ministry of Natural Resources and Energy. *Forest Management Manual for New Brunswick.* June 2004.

−. *Crown Licenses.*

−. *A Vision for New Brunswick's Forests.* March 2000.

−. *Objectives and Standards for the New Brunswick Crown Forest for the 2007-2012 Period.* 2005.

−. *Private Land Stewardship in New Brunswick: A Guide for Landowners.*

−. *Forest Operations Compliance Audit Performance Indicators.* June 2007.

−. *Licensee Performance Evaluation 1997-2002.* Updated February 2006.

−. *Forest Audit Reports.*

Jaako Poyry Consulting. *New Brunswick's Crown Forests: Assessment of Stewardship and Management.* November 2002.

Nova Scotia

Province of Nova Scotia. *Crown Lands Act.* R.S., c. 114, s. 1.

–. *Forests Act.* R.S., c. 179, s. 1.

–. *Forest Sustainability Regulation.* N.S. reg. 284/207.

–. *Scott Maritimes Limited Agreement (1965) Act.* R.S., c. 115, s. 1.

–. *Stora Forest Industries Agreement Act.* R.S., c. 446, s. 1.

Nova Scotia Department of Natural Resources. *Code of Forest Practice – Interim Guidelines for Crown Land.* April 2008.

–. *Stumpage Indexing and Annual Adjustment Methods.* 18 March 2005.

–. *Review and Recommendations on the Valuation, Allocation and Sale of Crown Timber Resources in Nova Scotia.* AgFor Incorporated, Fredericton, NB. December 2000.

National Aboriginal Forestry Association. *Aboriginal-Held Forest Tenures in Canada, 2002-2003.* 2003.

Newfoundland and Labrador

Province of Newfoundland and Labrador. *Forest Act* [R.S.N.L. 1990] F-23.

–. *Cutting of Timber Regulation.* C.N.L.R. 1108/96.

–. *Timber Royalty Regulation.* C.N.L.R. 962/96.

–. *Forest Protection Act* [R.S.N.L.] F-22.

Newfoundland and Labrador Department of Natural Resources. Forest Services Branch. *Provincial Sustainable Forest. Management Strategy.* 2003.

–. *Permits Licences and Fees.*

–. *Forest Management History.*

Notes

Introduction

1 In 1990, the Canadian Forest Service published a description and comparison of Canadian forest policies, as administered through provincial Crown forest tenure systems (Haley and Luckert 1990). Although limited in scope and now outdated, it remains the only published country-wide comparison of Crown forest tenure arrangements.

2 Although Crown title to the land was proclaimed at the time of colonization, more recent events suggest that the presumption of Crown title is questionable. Since 1982, when "existing Aboriginal and treaty rights" were affirmed in Canada's Constitution Act, the relationship between First Nations people and governments in Canada has been dramatically transformed. The debate has been centred in British Columbia, but the jurisprudence surrounding Aboriginal rights and title originating in that province has had a profound impact on the rest of Canada. Across Canada, Aboriginal people are gaining recognition that substantial areas of their traditional territories may remain exclusively in their hands. Land claims negotiations are underway in the territories and several provinces. These events will undoubtedly play an important role in shaping both federal and provincial forest polices in the future (National Round Table on the Environment and the Economy 2005).

3 30 & 31 Victoria, c. 3. (U.K.).

4 *Canada Act 1982* (U.K.), 1982, c. 11.

5 For example, a 1944 study by the BC Bureau of Economics and Statistics identified thirteen small towns in the East Kootenay District of British Columbia alone that had become ghost towns or entered into periods of serious economic decline

as a result of reduced forest sector activity during the preceding twenty-five years (Mercer 1944).

6 For further discussion of the interpretation and implications of sustainable forest management, see Haley and Luckert (1995) or Burton et al. (2003).

7 During the period from September 2004 to March 2006, there were 62 mill closures across Canada (14 in 2004-05 and 46 in 2005-06), including 6 pulp-and-paper mills and a lumber mill in the Atlantic provinces; 6 pulp-and-paper and 6 lumber mills in Quebec; 9 pulp-and-paper and 7 lumber mills in Ontario; 3 lumber mills in Alberta; and 2 pulp-and-paper and 4 lumber mills in British Columbia (Canadian Forest Service 2005, 2006). The rate of closures accelerated during the period 2006-07 (ForestTalk.com 2008).

8 An exception is British Columbia, where over the past three years fundamental policy reforms have been made, with the objective of making the forest products sector more competitive by lowering costs and providing firms with greater flexibility to pursue profit-maximizing strategies and respond to changing market conditions.

9 Personal communication with Kirk Andries, Ursus, 5 May 2008.

Chapter 1: The Rise of Sustainable Forest Management and Trends in Forest Sector Governance

1 For a comprehensive portrayal of trends in the governance of forest resources internationally, see Gluck et al. (2005).

2 J. Wilson (2002) provides a general Canadian overview, but it is not focused explicitly on the forest sector. His earlier work (J. Wilson 1998) provides a thorough analysis of the forest-environmental movement in British Columbia, but similar work on other provinces does not exist.

3 For a thorough examination of Canada's forest policies by one of Canada's leading environmentalists, see E. May (2005).

4 For an example of an advocacy research report, see the report on the mountain caribou by a coalition of environmental groups (Page, Scott, and Batycki 2005).

5 The Great Bear Rainforest campaign is described by Howlett, Rayner, and Tollefson 2009a). Information on the boreal campaigns is accessible from http://www.forestethics.org and http://www.greenpeace.ca.

6 *Delgamuukw v. The Queen*, [1997] 3 S.C.R. 1010.

7 *Haida Nation v. British Columbia (Ministry of Forests)* 2004 SCC 73 (18 November 2004); *Taku River Tlingit Nation v. British Columbia (Environmental Assessment Office)* 2004 SCC 74.

8 *Bernard v. The Queen*, 2003 NBCA 55.

9 Respectively, R.S.B.C. 1996, c. 159; S.O. 1994, c. 25; Alberta Sustainable Resource Development (2008).

10 Forest and Range Practices Act, S.B.C. 2002, c. 69.

11 For the development of a concept applicable to land use planning, see Weber and Adamowicz (2002).

12 The Government of Ontario announced in April 2004 that it will "require that all Sustainable Forest Licence (SFL) holders be certified to an accepted performance standard by the end of 2007." See http://www.mnr.gov.on.ca/.

13 Benjamin Cashore and colleagues refer to forest certification as "governing through markets" (Cashore, Auld, and Newsom 2004).

14 For an elaboration of the political logic behind jurisdictional struggles in Canadian federalism, see Harrison (1996).

Chapter 2: A General Framework for a Comparative Analysis of Canadian Crown Forest Tenures

1 See, for example, the International Association for the Study of the Commons at http://www.indiana.edu/~iascp/.

2 Utilization standards regulate the size and quality of trees and logs that must be harvested and removed from the forest following logging. They generally specify minimum tree diameters, maximum stump heights, minimum log sizes, and the proportion of rot and decayed wood a log must contain before it can be left in the forest. Departures from these standards may attract substantial penalties.

Chapter 3: Crown Forest Tenures in Canada

1 Interim measures agreements between the BC government and First Nations provide for the protection, management, or use of land and resources before treaties are concluded. The agreements are designed to deliver immediate benefits to First Nations, serve as building blocks for final treaties, and provide a greater degree of certainty for land management and for business development.

2 Forest Act R.S.B.C., c. 157, Division 8.1.

3 Forest Act R.S.B.C., c. 157, s. 43.3(c).

4 In Quebec, contrats d'approvisionnement et d'aménagement forestier (CAAFs) may be transferred as collateral in consideration of a loan or line of credit.

5 Forest Act R.S.B.C., c. 157, s. 54.1.

6 In December 2008, in the wake of AbitibiBowater's announcement that it was permanently closing its newsprint mill at Grand Falls-Windsor, the premier of Newfoundland and Labrador announced that his government intended to expropriate the company's timber harvesting, water, and hydroelectric power rights.

7 "Replaceability" differs from "renewability" in that when a licence is replaced, a completely new licence is issued, whereas when a licence is renewed, the existing licence is simply extended for an additional term. Replaceability is said to offer governments more flexibility than renewability in adjusting licences to meet changing social attitudes and objectives.

8   Public goods are non-exclusive and non-rival. Non-exclusive means that once a good or service is provided, it is not possible to exclude anyone from making use of it. Non-rival means that the enjoyment user of a good or service receives is not influenced by the number of people who enjoy the public good. A classical example of a public good is a lighthouse; once it is built, no mariner can be excluded from using it, and it doesn't matter to any user how many others are benefiting from the beacon. These attributes make it difficult, if not impossible, for markets to supply such services, as firms do not have incentives to supply goods or services from which they cannot exclude people and make them pay for them. Biodiversity is an example of a forest product that is a public good. Sometimes a product has the characteristics of a public good if, theoretically, access could be controlled but it is difficult to do so for economic reasons; that is, the costs of controlling access and administering user fees exceed the value of the revenue collected. Such goods, however, can become marketable goods as demand and potential revenues rise. Forest sector examples include certain types of recreation and the harvesting of non-timber botanical products.

9   Recognition of market and political uncertainties when making investment decisions ensures that firms don't choose strategies based on unrealistic expectations.

10   The difficulty in providing appropriate incentives in the presence of compensation has been referred to as the "paradox of compensation" (e.g., Cooter and Ulen 2000).

Chapter 4: Regulating for Sustainable Forest Management

Research assistance for this chapter was provided by Ravi Hegde, Jennifer Karmona, and Peter Wood.

1   For a discussion of this change, see Golec and Luckert (2008). Although CSA-Z809-02 requirements must be addressed in the plan, a separate third-party verification audit must be passed in order to achieve Canadian Standards Association certification.

2   Crown Forest Sustainability Act, 1994, S.O. 1994, c. 25.

3   The two most recent comprehensive reviews of BC forest policy recommended shifting volume-based licences to area-based ones (Forest Resources Commission 1991; BC Forest Policy Review 2000).

4   B.C. Reg. 14/2004.

5   Forest and Range Practices Act, S.B.C. 2002, c. 69.

Chapter 5: Interprovincial Comparison of Crown Stumpage Fee Systems

Research assistance for this chapter was provided by Chris Arnot and Rong Lu.

1   Nautiyal (1988) noted that the terms "stumpage" and "stumpage fees" are frequently used interchangeably. For the purposes of this book, the convention adopted by Nautiyal is followed in that "stumpage" is used to refer to physical

standing inventories of timber, whereas "stumpage fees" is used to describe the amount the owner of the timber receives when rights to harvest standing trees are sold. As will become clear as this chapter progresses, stumpage fees may or may not reflect the true value of standing timber depending on how they are calculated and/or collected.

2 For example, PricewaterhouseCoopers and International Wood Markets Research Inc. (2003) estimated that stumpage fees accounted for approximately 25 percent of the delivered wood costs for firms in the Canadian Prairie provinces in 2002.

3 For a description and analysis of the countervailing duty case, see Percy and Yoder (1987); K. van Kooten (2002), and Zhang (2007).

4 Although standing trees often embody non-timber values as well, the following discussion concentrates on the fibre value of trees as they may be used to make timber products.

5 Opportunity cost is a measure of the return that a productive input could earn in its next best use.

6 In economics, a "normal" profit is regarded as a cost – the return to entrepreneurs for taking risks and using their skills to bring together and organize the various productive inputs. There is no hard and fast rule as to what a "normal" profit should be, since it will vary with degree of risk and the opportunity costs of entrepreneurs. The best we can say is that a normal profit is that which is necessary to attract entrepreneurs to an enterprise and retain them.

7 Luckert (1991a) has developed a method that illustrates the importance of tenure restrictions in influencing the costs of forestry firms.

8 The costs associated with changing forest policies have been referred to as the "dynamic costs of regulation" (Boyd and Hyde 1989).

9 For a discussion of bilateral monopoly behaviour, see, for example, Stigler (1966).

10 These situations are referred to by economists as "monopsony" where there is only one buyer, and "oligopsony" where there are a small number of buyers.

11 Prior to 2005, when it was discontinued, British Columbia had a program – the Small Business Forest Enterprise Program (SBFEP) – under which about 12 percent of the allowable annual cut was reserved for small businesses that did not hold major Crown forest tenures. This volume was sold competitively in the form of small, short-term timber sale licences that reverted to the Crown once harvesting was completed.

12 Important to an understanding of this phenomenon is the concept of the economic margin. A log within the economic margin is one that costs less to harvest than it is worth in the marketplace. Such a log when harvested contributes to economic rent. A submarginal log is one that costs more to harvest than it is worth. Such a log, if harvested, reduces the economic rent. A fixed stumpage fee charged per cubic metre harvested adds to the cost of harvesting each log; consequently, some logs that would be within the economic margin without the fee become

submarginal when the fee is applied. Thus, a rational firm will not harvest these logs and the total volume harvested will decrease.

Chapter 6: In Search of Forest Tenures for Sustainable Forest Management

1 A study by Cashore and McDermott (2004) (see Chapter 4) compares forest practices in thirty-six jurisdictions around the world and concluded that the Canadian provinces included – British Columbia, Alberta, Ontario, and Quebec – have more stringent regulatory regimes to protect environmental values than other regions.

2 Indeed, only about 60 percent of Canada's forestland is considered to be timber productive, and of this a considerable area cannot be harvested economically (Canadian Council of Forest Ministers, National Forest Data Base, http://nfdp.ccfm.org/).

3 Schweitzer, Sassaman, and Schallau (1972, 416) defined the allowable cut effect as an "immediate increase in today's allowable cut which is attributable to expected future increases in [timber] yields."

4 There are exceptions to this generalization on some of the better sites in British Columbia. For example, Luckert and Haley (1990) and Zhang and Pearse (1996) found investments in silviculture being made on private lands that exceeded subsidized investments on public land.

5 For example, there has been very little monitoring of the community forests program in British Columbia that began as a small number of experimental pilot projects (Ambus 2008).

6 These concepts underlay the "paradox of compensation" that is discussed at length in the economics and law literature (e.g., Cooter and Ulen 2000).

7 Gunningham and Sinclair (1999) refer to "smart" regulations as policies that better meet objectives through the use of multiple policy instruments and a broad range of regulatory actors.

References

Alberta Sustainable Resource Development. 2006a. *Alberta Forest Management Planning Standard*. April 2006. http://srd.alberta.ca/forests/.

–. 2006b. *Alberta Timber Harvest Planning and Operating Ground Rules Framework for Renewal*. December. http://srd.alberta.ca/.

–. 2008. *Alberta Timber Harvest Planning and Operating Ground Rules Framework for Renewal*. January. http://www.srd.gov.ab.ca/.

Alchian, A.A., and H. Demetz. 1973. The Property Rights Paradigm. *Journal of Economic History* 33(1): 16-27.

Ambus, L.M. 2008. *The Evolution of Devolution: Evaluation of Community Forest Agreements in British Columbia*. MSc thesis, Department of Forest Resources Management, University of British Columbia.

Anderson, J.A. 2008. *Economics of Priority Use Zoning*. PhD diss., Department of Rural Economy, University of Alberta.

Anderson, J.A., and M.K. Luckert. 2007. Can Hybrid Poplar Save Industrial Forestry in Canada? A Financial Analysis in Alberta and Policy Implications. *Forestry Chronicle* 38(1): 92-104.

Association of BC Forest Professionals. 2005. *Forest Legislation and Policy Reference Guide 2005*. Vancouver: ABCFP. http://www.abcfp.ca/.

BC Competition Council. 2006a. *Pulp and Paper Industry Advisory Committee: Final Report*. Victoria: Government of British Columbia. http://www.bccompetition council.gov.bc.ca/.

–. 2006b. *Wood Products Industry Advisory Committee: Report to the Council, March 2006*. Victoria: Government of British Columbia. http://www.bccompetition council.gov.bc.ca/.

BC Forest Policy Review. 2000. *Shaping Our Future*. Victoria: Government of British Columbia. http://www.for.gov.bc.ca/.

BC Forest Practices Board. 2004. *Implementation of Biodiversity Measures under the Forest Practices Code Implications for the Transition to the Forest and Range Practices Act Special Report*. Victoria: Government of British Columbia. http://www.fpb.gov.bc.ca/.

–. 2006. *A Review of the Early Forest Stewardship Plans Under FRPA, Special Report*. FPB/SR/28, May. http://www.fpb.gov.bc.ca/.

BC Ministry of Agriculture and Lands. 2006. *Report on the Status of Strategic Land Use Plans in British Columbia*. Integrated Land Management Bureau, 31 October. http://www.agf.gov.bc.ca/.

BC Ministry of Forests and Range. 2008. *Forestry Round Table Focuses on the Future*. Press release 2008FOR0028-000326.

BC Office of the Premier. 2005. *Premier Appoints New Asia-Pacific Competition Councils*. Press release 2005OTP47-000372.

Benson, C.A. 1988. A Need for Extensive Forest Management. *Forestry Chronicle* 63: 421-30.

Bernstein, S., and B. Cashore. 2000. Globalization, Four Paths of Internationalization and Domestic Policy Change: The Case of Ecoforestry in British Columbia. *Canadian Journal of Political Science* 33(1): 67-99.

Blume, L., D. Rubinfield, and P. Shapiro. 1984. The Taking of Land: When Should Compensation Be Paid? *Quarterly Journal of Economics* 100: 71-92.

Bouthillier, L., D. Chua, B. Laplante, and M.K. Luckert. 1992. Reflexion economique sur le rendement soutenu, le développement durable et l'aménagement des ressources forestières. Cahiers de recherche. Working Paper no. 92-13. Department of Economics, Laval University.

Boyd, R.G., and W.F. Hyde. 1989. *Forest Sector Intervention: The Impacts of Regulation on Social Welfare*. Ames: Iowa State University Press.

Bromley, D. 1991. *Environment and Economy: Property Rights and Public Policy*. Cambridge, MA: Basil Blackwell.

Brundtland, G.H. 1987. *Our Common Future*. Report of the United Nations' World Commission on Environment and Development. New York: United Nations.

Burton, P.J., C. Messier, D.W. Smith, and W.L. Adamowicz, eds. 2003. *Towards Sustainable Management of the Boreal Forest*. Ottawa: NRC Research Press.

Canadian Boreal Initiative. 2003. *Canadian Boreal Forest Conservation Framework*. http://www.borealcanada.ca/.

Canadian Forest Service. 2005. *The State of Canada's Forests 2004-2005*. Ottawa: Natural Resources Canada.

–. 2006. *The State of Canada's Forests 2005-2006*. Ottawa: Natural Resources Canada.

–. 2007. *The State of Canada's Forests 2006-2007*. Ottawa: Natural Resources Canada.

Cartwright, J. 2003. Environmental Groups, Ontario's Lands for Life Process and the Forest Accord. *Environmental Politics* 12(2): 115-32.

Cashore, B., and G. Auld. 2003. British Columbia's Environmental Forestry Policy Record in Perspective. *Journal of Forestry* 101(8): 42-47.

Cashore, B., G. Auld, and D. Newsom. 2004. *Governing through Markets: Forest Certification and the Emergence of Non-State Authority.* New Haven, CT: Yale University Press.

Cashore, B., and C. McDermott. 2004. *Global Environmental Forest Policies: Canada as a Constant Case Comparison of Select Forest Practice Regulations.* Victoria: International Forest Resources.

CCFM (Canadian Council of Forest Ministers). 1992. *Sustainable Forests: A Canadian Commitment.* Hull, QC: CCFM.

Ciriacy-Wantrup, S.V., and R.C. Bishop. 1975. "Common Property" as a Concept in Natural Resources Policy. *Natural Resources Journal* 15: 713-27.

Clarke, J. 1997. Logging Costs Skyrocketing in BC. *Logging and Sawmilling Journal,* May 1997. http://www.forestnet.com/achives/.

Coase, R.H. The Problem of Social Cost. *Journal of Law and Economics* 3: 1-44.

Coglianese, C., and D. Lazer. 2003. Management-Based Regulation: Prescribing Private Management to Achieve Public Goals. *Law and Society* 37: 691-730.

Cooter, R., and T. Ulen. 2000. *Law and Economics.* 3rd ed. Reading, MA: Addison-Wesley.

Crawford, S.E.S., and E. Ostrom. 1995. A Grammar of Institutions. *American Political Science Review* 89: 582-600.

Cumming, S.G., and G.W. Armstrong. 2001. Divided Land Base and Overlapping Tenure in Alberta, Canada: A Simulation Study Exploring Costs of Forest Policy. *Forestry Chronicle* 77: 501-08.

Dahlman, C. 1980. *The Open Field System and Beyond: A Property Rights Analysis of an Economic Institution.* Cambridge: Cambridge University Press.

Dales, J.H. 1968. *Pollution Property and Prices.* Toronto: University of Toronto Press.

Davis, L., K.N. Johnson, P. Bettinger, and T. Howard. 2000. *Forest Management.* 4th ed. New York: McGraw-Hill.

Diver, C. 1989. Regulatory Precision. In *Making Regulatory Policy.* Ed. K. Hawkins and J. Thomas, 199-232. Pittsburgh: University of Pittsburgh Press.

Dowdle, B. 1984. The Case for Selling Federal Timber Lands. In *Selling the Federal Forests.* Ed. A.E. Gamache, 21-46. Seattle: University of Washington Press.

FAO (Food and Agricultural Organization of the United Nations). 2005. The Global Forest Resources Assessment 2005. FAO Forestry Paper 147. Rome: FAO.

Fischer, I. 1923. *Elementary Principles of Economics.* New York: Macmillan. Cited in S. Pejovich, 1984, 164.

ForestTalk.com. 2008. *ForestTalk.com – Canada's Forestry Blog.* Blog entries by Lisa Schuyler. http://foresttalk.com/.

Forest Economics and Policy Forum. 2005. *Proceedings of Besieged by Global Change: Defining the Future of B.C.'s Forest Sector.* Vancouver: British Columbia Forest Economics and Policy Forum, University of British Columbia.

Forest Products Association of Canada. 2006. *Transformation – 2005 Annual Review.* Ottawa: Forest Products Association of Canada.

Forest Resources Commission. 1991. *The Future of Our Forests.* http://www.for. gov.bc.ca/.

Frame, T.M., T. Gunton, and J.C. Day. 2004. The Role of Collaboration in Environmental Management: An Evaluation of Land and Resource Planning in British Columbia. *Journal of Environmental Planning and Management* 47(1) (January 2004): 59-82.

Gillis, R.P., and T.R. Roach. 1986. *Lost Initiatives: Canada's Forest Industries, Forest Policy, and Forest Conservation.* New York: Greenwood Press.

Glastra, R., ed. 1999. *Cut and Run: Illegal Logging in the Tropics.* Ottawa: International Development Research Centre.

Gluck, P., et al. 2005. Change in the Governance of Forest Resources. In *Forests in the Global Balance – Changing Paradigms.* IUFRO World Series. Vol. 17. Ed. G. Mery, R. Alfaro, M. Kanninen, and M. Labovikov, 51-74. Helsinki: IUFRO.

Golec, P.J., and M.K. Luckert. 2008. Would Harmonizing Public Land Forest Policies, Criteria and Indicators, and Certification Improve Progress Towards Sustainable Forest Management? A Case Study in Alberta, Canada. *Forestry Chronicle* 84(3): 410-19.

Gunningham, N., and P. Grabowsky. 1998. *Smart Regulation.* Oxford: Clarendon Press.

Gunningham, N., and D. Sinclair. 1999. Designing Smart Regulation. http://www. oecd.org/.

Haley, D. 1996. Paying the Piper: The Cost of the British Columbia Forest Practices Code. *RPF Forum* 3(5): 26-28.

–. 2001. *Reforming the Crown Stumpage System in British Columbia. ABCPF Forum* 8(5), Association of BC Professional Foresters, Vancouver.

–. 2002. Community Forestry in British Columbia: The Past Is Prologue. *Trees, Forests and People* 46: 54-61.

–. 2003. *Are Log Export Restrictions on Private Forestland Good Public Policy?* Victoria: Private Forest Landowners Association.

–. 2006. *British Columbia's Crown Forest Tenure System in a Changing World: Challenges and Opportunities.* Paper presented at Creating New Opportunities: Forest Tenure and Land Management in British Columbia, a symposium of the BC Forum on Forest Economics and Policy, Vancouver.

Haley, D., and M.K. Luckert. 1990. *Forest Tenures in Canada: A Framework for Policy Analysis.* Information report E-X-43. Ottawa: Economics Directorate, Forestry Canada.

–. 1991. The Effects of Canadian Forest Tenures on the Organizational Structure and Capital Budgeting Procedures of Forestry Firms. Working Paper no. 167. Forest Economics and Policy Analysis Unit, University of British Columbia.

–. 1995. Policy Instruments for Sustainable Development in the British Columbia Forestry Sector. In *Managing Natural Resources in British Columbia: Markets, Regulations, and Sustainable Development.* Ed. Anthony Scott, John Robinson, and David Cohen, 54-79. Vancouver: UBC Press.

–. 1998. Tenures as Economic Instruments of Achieving Objectives of Public Forest Policy in British Columbia. In *The Wealth of Forests: Markets, Regulation, and Sustainable Development.* Ed. C. Tollefson, 123-51. Vancouver: UBC Press.

Haley, D., and H. Nelson. 2006. British Columbia's Crown Forest Tenure System in a Changing World: Challenges and Opportunities. The Forum on Conservation, Economics and Policy, Synthesis Paper, SP 06-01. University of British Columbia, Vancouver.

Hall, P., and R. Taylor. 1996. Political Science and the Three New Institutionalisms. *Political Studies* 44: 936-57.

Harrison, K. 1996. *Passing the Buck: Federalism and Environmental Policy.* Vancouver: UBC Press.

–. 2001. Voluntarism and Environmental Governance. In *Governing the Environment.* Ed. E. Parson, 207-46. Toronto: University of Toronto Press.

Hegan, L., and M.K. Luckert. 2000. An Economic Assessment of the Allowable Cut Effect (ACE) for Enhanced Forest Management Policies: An Alberta Case Study. *Canadian Journal of Forest Research* 30(10): 1591-600.

Helsinki Process. 1993. *General Guidelines for the Sustainable Management of Forests in Europe.* Final communiqué of the Pan-European Ministerial Conference on the Protection of Forests in Europe, Helsinki.

Hoberg, G. 2000. How the Way We Make Policy Governs the Policy We Choose. In *Forging Truces in the War in the Woods: Sustaining the Forests of the Pacific Northwest.* Ed. D. Alper and D. Salazar, 26-53. Vancouver: UBC Press.

–. 2001a. The 6 Percent Solution: The Forest Practices Code. In *In Search of Sustainability: BC Forest Policy in the 1990s.* Ed. Benjamin Cashore, George Hoberg, Michael Howlett, Jeremy Rayner, and Jeremy Wilson, 61-93. Vancouver: UBC Press.

–. 2001b. Policy Cycles and Policy Regimes: A Framework for Studying Policy Change. In *In Search of Sustainability: British Columbia Forest Policy in the 1990s.* Ed. Benjamin Cashore, George Hoberg, Michael Howlett, Jeremy Rayner, and Jeremy Wilson, 3-30. Vancouver: UBC Press.

–. 2002. *Finding the Right Balance Report of Stakeholder Consultations on a Results-Based Forest and Range Practices Regime for British Columbia.* 25 July 2002. http://www.for.gov.bc.ca/.

Hoberg, G., and J. Karmona. 2005. *Policies for Protecting Biodiversity in Canada's Forests – Divergent Approaches among the Provinces.* Paper presented at IUFRO World Congress, Brisbane.

Howlett, M. 1990. The Round Table Experience: Representation and Legitimacy in Canadian Environmental Policy. *Queen's Quarterly* 97: 580-601.

–, ed. 2001a. *Canadian Forest Policy: Adapting to Change.* Toronto: University of Toronto Press.

–. 2001b. The Federal Role in Canadian Forest Policy: From Territorial Landowner to International and Intergovernmental Co-ordinating Agency. In *Canadian Forest Policy: Adapting to Change.* Ed. M. Howlett, 378-415. Toronto: University of Toronto Press.

Howlett, M., and J. Rayner. 2001. The Business and Government Nexus: Principal Elements and Dynamics of the Canadian Forest Policy Regime. In *Canadian Forest Policy: Adapting to Change.* Ed. M. Howlett, 23-62. Toronto: University of Toronto Press.

–. 2006a. Convergence and Divergence in "New Governance" Arrangements: Evidence from European Integrated Resources Strategies. *Journal of Public Policy* 26: 167-89.

Howlett, M., J. Rayner, and C. Tollefson. 2009a. From Governance to Governance in Forest Planning? Lessons from the Case of the British Columbia Great Bear Rainforest Initiative. *Forest Policy and Economics* 11(5-6): 383-91.

–. 2009b. From Old to New Governance in Canadian Forest Policy. In *Canadian Environmental Policy and Politics: Prospects for Leadership and Innovation.* 3rd ed. Ed. D.L. VanNijnatten and R. Boardman, 183-96. Toronto: Oxford University Press.

Hyde, W.F., and R.A. Sedjo. 1992. Managing Tropical Forests: Reflections on the Rent Distribution Discussion. *Land Economics* 68: 343-50.

Innes, R. 1995. An Essay on Takings: Concepts and Issues. *Choices* 10(1): 4-7, 42-44.

International Tropical Timber Organization. 1998. *Criteria and Indicators for Sustainable Management of Natural Tropical Forests.* ITTO Policy Development Series no. 7. Yokohama, Japan.

Jepperson, R. 1991. Institutions, Institutional Effects, and Institutionalism. In *The New Institutionalism in Organizational Analysis.* Ed. W. Powell and P. DiMaggio, 143-63. Chicago: University of Chicago Press.

Kimmins, H. 2006. Ecosystem Tenures: Institutional Arrangements to Promote Stewardship and Sustainability. The Forum on Conversation, Economics and Policy, Synthesis Paper, SP 06-03. Faculty of Forestry, University of British Columbia, Vancouver.

Kimmins, J.P. 2000. Respect for Nature: An Essential Foundation for Sustainable Forest Management. In *Ecosystem Management of Forested Landscapes: Directions and Implications.* Ed. R.G. D'Eon, J.F. Johnson, and E.A. Ferguson, 3-24. Vancouver: UBC Press.

KPMG Global Sustainability Services. 2005. *KPMG International Survey of Corporate Responsibility Reporting 2005.* Amsterdam: KPMG Global Sustainability Services.

Larson, A., and J. Ribot. 2004. Democratic Decentralization through a Natural Resource Lens: An Introduction. *European Journal of Development Research* 16(1) (Spring 2004): 1-25.

Le Master, D.C., and J.H. Beuter, eds. 1989. *Community Stability and Forest-Based Economies.* Portland: Timber Press.

Lindquist, E., and A. Wellstead. 2001. Making Sense of Complexity: Advance and Gaps in Comprehending the Canadian Forest Policy Process. In *Canadian Forest Policy: Adapting to Change.* Ed. M. Howlett, 419-46. Toronto: University of Toronto Press.

Luckert, M.K. 1991a. Effect of Canadian Forest Tenures on Rent Distributions and Resource Allocations: A British Columbia Case Study. *Forest Science* 37(5): 1441-62.

–. 1991b. The Perceived Security of Institutional Investment Environments of Some British Columbia Forest Tenures. *Canadian Journal of Forest Research* 21: 318-25.

–. 1998. Efficiency Implications of Silvicultural Expenditures from Separating Ownership and Management on Forest Lands. *Forest Science* 44(3): 365-78.

–. 2001. Welfare Implications of the Allowable Cut Effect in the Context of Sustained Yield and Sustainable Development Forestry. *Journal of Forest Economics* 7(3): 203-24.

–. 2005. In Search of Optimal Institutions for Sustainable Forest Management. In *Institutions, Sustainability, and Natural Resources: Institutions for Sustainable Forest Management.* Ed. S. Kant and R.A. Berry, 21-42. Netherlands: Springer.

Luckert, M.K., and J.T. Bernard. 1993. What Is the Value of Standing Timber? Difficulties in Merging Theory with Reality. *Forestry Chronicle* 69(6): 680-85.

Luckert, M.K., and P. Boxall. 2009. Institutional Vacuums in Canadian Forest Policy: Can Criteria and Indicators and Certification of Sustainable Forest Management Fill the Void? *Forestry Chronicle* 85(2): 277-84.

Luckert, M.K., and D. Haley. 1990. The Implications of Various Silvicultural Funding Arrangements for Privately Managed Public Forest Land in Canada. *New Forests* 4(1): 1-12.

–. 1993. Canadian Forest Tenures and the Silvicultural Investment Behavior of Rational Firms. *Canadian Journal of Forest Research* 23: 1060-64.

Luckert, M.K., and T.B. Williamson. 2005. Should Sustained Yield Be Part of Sustainable Forest Management? *Canadian Journal of Forest Research* 35(2): 356-64.

Martin, Gwen. 2003. *Management of New Brunswick's Crown Forests*. N.p.: New Brunswick Department of Natural Resources. http://www.gnb.ca/0079/pdf/managing_NB_crown_forests-e.pdf.

Mascarenhas, M., and R. Scarce. 2004. "The Intention Was Good": Legitimacy, Consensus-Based Decision Making, and the Case of Forest Planning in British Columbia, Canada. *Society and Natural Resources* 17 (2004): 17-38.

May, E. 2005. *At the Cutting Edge: The Crisis in Canada's Forests*. Rev. and expanded edition. Toronto: Key Porter.

May, P. 2007. Regulatory Regimes and Accountability. *Regulation and Governance* 1: 8-26.

McCarthy, J. 2006. Neo-Liberalism and the Politics of Alternatives: Community Forestry in British Columbia and the United States. *Annals of the Association of American Geographers* 96: 84-104.

McDermott, C.L., and G. Hoberg. 2003. From State to Market: Forestry Certification in the U.S. and Canada. In *Two Paths toward Sustainable Forests: Public Values in Canada and the United States*. Ed. B. Schindler, T. Beckley, and C. Finley, 229-50. Corvallis: Oregon State University Press.

Mercer, W.M. 1944. *Growth of Ghost Towns*. Bureau of Economics and Statistics, BC Department of Trade and Commerce. Victoria: King's Printer.

Minister's Council on Forest Sector Competitiveness. 2005. *Final Report of the Minister's Council on Forest Sector Competitiveness, November 2005*. Toronto: Ontario Ministry of Natural Resources.

MRNFP (Ministère des Resources naturelles, de la Faune et des Parcs. 2004. *Forest Resource Protection and Development Objectives: General Forest Management Plans 2007-2012 Implementation Document*. Quebec: Government of Quebec.

Murphy, P.J., R. Udell, B. Bott, and R.E. Stevenson. 2002. *The Hinton Forest 1955-2000: A Case Study in Adaptive Forest Management*. Foothills Model Forest History Series. Vol. 2. Hinton, AB: Foothills Research Institute.

National Forest Strategy Coalition. 1992. *Canada Forest Accord*. Ottawa: National Forest Strategy Coalition.

National Round Table on the Environment and the Economy. 2005. *Aboriginal Issues in Canada's Boreal Forest*. Boreal Forest Program. Ottawa: National Round Table on the Environment and the Economy.

Natural Resources Canada. 2007. *The State of Canada's Forests: Annual Report 2007*. http://foretscanada.rncan.gc.ca/rpt.

Nautiyal, J.C. 1980. An Anatomy of Stumpage Value. *Canadian Journal of Forest Research* 10: 135-42.

–. 1988. *Forest Economics: Principles and Applications*. Toronto: Canadian Scholars' Press.

Nautiyal, J.C., and D.V. Love. 1971. Some Economic Implications of Methods of Charging Stumpage. *Forestry Chronicle* 47(1): 25-28.

Nelson, J.D. 2004. Forest Level Planning. In *Forestry Handbook for British Columbia*. 5th ed. Ed. S.B. Watts and L. Tolland, 25-45. Vancouver: University of British Columbia Forestry Undergraduate Society.

New Brunswick Department of Natural Resources. 2006. *Licensee Performance Evaluation 1997-2002* (updated February 2006). http://www.gnb.ca/.

North, D.C. 1993. *The New Institutional Economics and Development*. EconPapers Economics at Your Fingertips, http://129.3.20.41/eps/eh/papers/9309/9309002.pdf.

Ontario Ministry of Natural Resources. 1995. *Forest Operations and Silviculture Manual*. 1st ed. Toronto: Ontario Ministry of Natural Resources.

–. 2009. *Forest Management Planning Manual for Ontario's Crown Forests*. http://www.mnr.gov.on.ca/.

Ostrom, E. 1990. *Governing the Commons*. Cambridge: Cambridge University Press.

–. 1999. Institutional Rational Choice: An Assessment of the Institutional Analysis and Development Framework. In *Theories of the Policy Process*. Ed. P.A. Sabatier, 35-71. Boulder, CO: Westview Press.

Page, D., J. Scott, and C. Batycki. 2005. *Staring at Extinction: Mountain Caribou in British Columbia; An Analysis of Planned Logging in BC's Inland Temperate Rainforest*. Mountain Caribou Project, http://www.mountaincaribou.org/pdf/staringatextinction_lores.pdf.

Pearse, P. 1976. *Timber Rights and Forest Policy in British Columbia, Report of the Royal Commission on Forest Resources, Volume 1*. Victoria: Queen's Printer.

–. 1988. Property Rights and the Development of Natural Resource Policies in Canada. *Canadian Public Policy* 14(3): 307-20.

–. 1992. *Evolution of the Forest Tenure System in British Columbia*. Victoria: British Columbia Ministry of Forests.

–. 1998. Economic Instruments for Promoting Sustainable Forestry: Opportunities and Constraints. In *The Wealth of Forests: Market, Regulation, and Sustainable Forestry*. Ed. C. Tollefson, 19-41. Vancouver: UBC Press.

–. 2000. *The Species at Risk Act and the Compensation Issue*. Environment Canada, http://www.ec.gc.ca/press/001221_m_e.htm.

–. 2001. *Ready for Change: Crisis and Opportunity in the Coast Forest Industry: A Report on the BC Forest Industry and Policy*. Report to the BC Minister of Forests, the Honourable Michael de Jong, November. http://www.for.gov.bc.ca/.

Pejovich. S. 1984. Origins and Consequences of Alternative Property Rights. In *Selling the Federal Forests*. Ed. A.E. Gameche, 163-75. Seattle: College of Forest Resources, University of Washington.

Percy, M.B., and C. Yoder. 1987. *The Softwood Lumber Dispute and Canada-U.S. Trade in Natural Resources*. Montreal: Institute for Research on Public Policy.

Pierre, J., and B.G. Peters. 2000. *Governance, Politics, and the State*. New York: St. Martin's Press.

Pierson, P. 2004. *Politics in Time*. Princeton, NJ: Princeton University Press.

PricewaterhouseCoopers and International Wood Markets Research Inc. 2003. *Global Lumber/Sawn Wood Cost Benchmarking Report.* http://www.pwc.com/.

Quebec Ministry of Natural Resources, Wildlife and Parks. 2005. *Forest Resource Protection and Development Objectives 2007-2012*. http://www.mrn.gouv.qc.ca/.

–. 2007. *Forest Planning (General and Annual Management Plans)*. http://www.mrn.gouv.qc.ca/.

Random Lengths. 2008. Random Lengths Framing Lumber Composite Price – by Month. http://www.randomlengths.com/.

Ravenel, Ramsay M., Ilmi M.E. Granoff, and Carrie A. Magee. 2004. Illegal Logging in the Tropics: A Synthesis of Issues. *Journal of Sustainable Forestry* 19(1-3): 351-71.

Rayner, J., and M. Howlett. 2007. The National Forest Strategy in Comparative Perspective. *Forestry Chronicle* 83(5): 651-57.

Roberts, D. 2008. Adapting to Changes in Values Due to Changes in Technology: Bioenergy. Seminar presented at Future Drivers of Tenure workshop, Vancouver: CIBC World Markets. http://www.sfmnetwork.ca/.

Rodriguez, P.M.J., M.K. Luckert, V. Lieffers, and V. Adamowicz. 1998. *Economic Analysis of Ecologically Based Mixedwood Silviculture at the Stand Level.* Project report to the Environmental Protection and Enhancement Fund, Alberta Environmental Protection, Edmonton.

Ross, M. 1995. *Forest Management in Canada*. Calgary: Canadian Institute of Resources Law.

Ross, M., and P. Smith. 2002. *Accommodation of Aboriginal Rights: The Need for an Aboriginal Forest Tenure.* Edmonton: Sustainable Forest Management Network.

Saastamoinen, O. 1999. Forest Policies, Access Rights and Non-Wood Forest Products in Northern Europe. *Unasylva* 50(3). http://www.fao.org/.

Sahajananthan S., D. Haley, and J. Nelson. 1998. Planning for Sustainable Forestry in B.C. through Land Use Zoning. *Canadian Public Policy* 24 (special supplement, May 1998): 73-82.

Salamon, L. 2002. The New Governance and the Tools of Public Action: An Introduction. In *The Tools of Government: A Guide to the New Governance*. Ed. L. Salamon, 1-47. Oxford: Oxford University Press.

Scott, A.D., and J. Johnson. 1983. *Property Rights: Developing the Characteristics of Interests in Natural Resources*. Discussion Paper no. 84-26. Department of Economics, University of British Columbia.

Schweitzer, D.L.R., R.W. Sassaman, and C.W. Schallau. 1972. Allowable Cut Effect: Some Physical and Economic Implications. *Journal of Forestry* 70: 415-18.

Stanbury, W.T. 2000. *Environmental Groups and the International Conflict over the Forests of British Columbia, 1990-2000.* Vancouver: Phelps Centre for the Study of Government and Business, Sauder School of Business, University of British Columbia.

Stanbury, W.T., and I. Vertinsky. 1998. Governing Instruments for Forest Policy in British Columbia: A Positive and Normative Analysis. In *The Wealth of Forests: Market, Regulation, and Sustainable Forestry*. Ed. C. Tollefson, 42-77. Vancouver: UBC Press.

Standing Committee on Forestry and Fisheries. 1990. *Forests of Canada: The Federal Role*. Ottawa: House of Commons.

Stavins, R.N. 2007. *A U.S. Cap and Trade System to Address Global Climate Change*. Washington, DC: Brookings Institution.

Stedman, R.C., R. Parkinson, and T.M. Beckley. 2005. Forest Dependency and Community Well-Being in Rural Canada: Variation by Forest Sector and Region. *Canadian Journal of Forest Research* 35: 215-20.

Stigler, G.J. 1966. *The Theory of Price*. 3rd ed. New York: Macmillan.

Sullivan, L. 1896. The Tall Office Building Artistically Considered. *Lippincott's Magazine*, March 1896. http://academics.triton.edu/faculty/fheitzman/talloffice building.html.

Thielmann, T., and C. Tollefson. 2009. Tears from an Onion: Layering, Exhaustion and Conversion in British Columbia Land Use Planning Policy. *Politics and Society* 28: 111-24.

Tollefson, C., F. Gale, and D. Haley. 2009. *Setting the Standards: Certification, Governance, and the Forest Stewardship Council*. Vancouver: UBC Press.

Uhler, R.S., and P.D. Morrison. 1986. *Utilization Standards and Economic Efficiency in British Columbia Forests*. Information Report 86-1. Forest Economics and Policy Analysis Project, University of British Columbia.

van Kooten, G.C. 1994. *Cost-Benefit Analysis of BC's Proposed Forest Practices Code*. Vancouver: BC Council of Forest Industries.

van Kooten, K. 2002. Economic Analysis of the Canada-United States Softwood Lumber Dispute: Playing the Quota Game. *Forest Science* 48(4): 712-21.

Vincent, J.R. 1990. Rent Capture and the Feasibility of Tropical Forest Management. *Land Economics* 66: 212-23.

–. 1993. Managing Tropical Forests: Comment. *Land Economics* 69: 313-18.

Vincent, J.R., and C.S. Binkley. 1993. Efficient Multiple-Use Forestry May Require Land-Use Specialization. *Land Economics* 69: 370-76.

Weber, M., and W. Adamowicz. 2002. Tradable Land-Use Rights for Cumulative Environmental Effects Management. *Canadian Public Policy* 28(4): 581-95.

Weimer, D., and A. Vining. 2005. *Policy Analysis: Concepts and Practice*. 4th ed. Upper Saddle River, NJ: Pearson Prentice Hall.

Wilson, J. 1998. *Talk and Log: Wilderness Politics in British Columbia*. Vancouver: UBC Press.

–. 2001. Talking the Talk and Walking the Walk: Reflections on the Early Influence of Ecosystem Management Ideas. In *Canadian Forest Policy: Adapting to Change*. Ed. M. Howlett, 94-126. Toronto: University of Toronto Press.

–. 2002. Continuity and Change in the Canadian Environmental Movement: Assessing the Effects of Institutionalization. In *Canadian Environmental Policy and Politics: Prospects for Leadership and Innovation*. 2nd ed. Ed. R. Boardman and D.L. VanNijnatten, 46-65. Toronto: Oxford University Press.

Wilson, J.Q., ed. 1980. *The Politics of Regulation*. New York: Basic Books.

Wilson, J., and J. Graham. 2005. *Relationship between First Nations and the Forest Industry: The Legal and Policy Context*. A report for the National Aboriginal Forestry Association, the Forest Products Association of Canada, and the First Nations' Forestry Program. Ottawa: Institute of Governance.

Winfield, M., and H. Benevides. 2003. *Industry Self-Inspection and Compliance in the Ontario Forest Sector*. Drayton Valley, AB: Pembina Institute for Appropriate Development.

Wolf, C. 1979. A Theory of Non-Market Failures. *Journal of Law and Economics* 22(1): 107-39.

–. 1988. *Markets or Governments: Choosing between Imperfect Alternatives*. Cambridge, MA: Massachusetts Institute of Technology.

Wynn, Graeme. 1980. *Timber Colony: An Historical Geography of Early Nineteenth Century New Brunswick*. Toronto: University of Toronto Press.

Zhang, D. 2007. *The Softwood Lumber War*. Washington, DC: Resources for the Future.

Zhang, D., and P.H. Pearse. 1996. Differences in Silvicultural Investment under Various Types of Forest Tenure in British Columbia. *Forest Science* 44(4): 442-49.

Index

References to tables or figures are indicated by "t" or "f" following the page number.

economic downturn (2008). *See* recession of 2008
economic margin, 187-88n12
economic rent, 116-19, 118f, 122, 143-44. *See also* stumpage fee systems
economic sustainability: challenges to, 151-53; under command-and-control policies, 162-64; defined, 83; export restrictions and, 91-92; forest planning and practice issues, 97-98, 101, 109, 112-13; of growing rights, 157; initial allocation of tenure and, 83-84; mill appurtenancy issues, 93-94; operational planning and, 111; policies for, and tradeoffs with other objectives, 150; stumpage systems and, 132-35; in sustainable forest management, 10-11; of sustained yield paradigm, 153-55; tenure comprehensiveness issues, 85, 86-87; tenure duration and renewability issues, 92-93; tenure mutability and compensation issues, 94-95; tenure size and, 89-90; transferability issues and frozen assets, 90-91; utilization standards and, 140-41; weak vs strong sustainability, 141-42
ecosystem diversity. *See* biodiversity
employment, in forest industry, 2, 9, 11, 91. *See also* economic sustainability; mill appurtenancy requirements; social sustainability
enforcement, of property rights, 48-49
enforcement and compliance, of tenure requirements, 64, 101, 109-10
environmental movement, 9-10, 19, 21-26, 22-23t, 39
environmental report cards, 24-25, 25t

environmental sustainability: under command-and-control policies, 162-63; defined, 82-83; forest planning and practice issues, 97-98, 101, 109, 112-13; forest productivity and, 151, 188n2; mill appurtenancy issues, 93; old-growth forests vs plantations, 151; policies for, and tradeoffs with other objectives, 149-50; protection and development objectives, 103-4; and protection of AAC, in BC, 155; in sustainable forest management, 11; tenure comprehensiveness issues, 85-86; tenure duration and renewability issues, 92-93; tenure exclusiveness issues, 90; tenure mutability and compensation issues, 95; tenure size issues, 88-90; utilization standards and, 140-41; volume-based allotments and, 88
l'Erreur Boréale, 23t
"evergreen" tenure renewals, 80
exclusiveness, of Crown tenure rights, 56; defined, 58t; as dimension of Crown tenure, 55-57, 55f, 61; interaction with private-public spectrum, 56-57, 186n8; provincial comparison, 76; and sustainable forest management, 90, 95
export dependence, 37
export restrictions, 58t, 62, 76-77, 77t, 91-92, 96. *See also* trade relations

federal government (Canada): Constitution Act (1867), 4-5; Constitution Act (1982), 5, 183n2; Fisheries Act, 169; forestland ownership and jurisdiction, 2-3, 3f, 5, 6t, 37; mismatch with provincial tenure

Printed and bound in Canada by Friesens

Set in Myriad and Sabon by Artegraphica Design Co. Ltd.

Copy Editor: Judy Phillips

Proofreader: Jillian Shoichet

Indexer: Judith Anderson